Conway Lloyd Morgan

Thomas Manss & Company
designers, narrators,
myth-makers, fabulators
and tellers of tales

aveditionrockets

4 **narrative** Introduction
6 **character** Bowers & Wilkins
26 **passion** Oberhavel Verkehrsgesellschaft
38 **desire** Meoclinic
50 **elegance** Oyuna
60 **continuity** Foster + Partners
76 **responsibility** Tate
88 **articulation** Vitsœ
102 **crossing borders** National Portrait Gallery
116 **telling stories** Salon
128 **simplicity** Internationales Design Zentrum
138 **vocabulary** Books and Magazines
150 **wit** Symbols and Logotypes

&

narrative

Design, it is often said, is all about the visual. This is true in the sense that visual criteria are most often used to evaluate design, and that design is mainly taken in by the eye (there are those who would argue that the eye is often taken in by design as well). But to emphasise the visual quality of design is to miss another important aspect. Design is not just about the visual, it is also about narrative. Thomas Manss & Company are designers, but they are also – and more importantly – narrators, myth-makers, fabulators, tellers of tales. Why is this so? Because design is a language, is about communication, and any communication is implicitly a narrative.

The more complex the communication, the more complex the narrative. Narrative does not have here only the simple sense of a story, a sequence of related events. It is also used in the sense of connectedness, of a statement only being comprehensible because its context is understood. This is a direct extension from Saussure's original thoughts about the linear and arbitrary nature of language. Just as every spoken statement occurs at a specific point in time and so is understandable only within the linguistic context of that time, so written statements (including ones written in a visual language) are read as contextual and contemporary. And just as language is synchronistically invariable and diachronistically variable, new visual statements can be understood within the contemporary context of visual language, but can as well extend it. An ampersand turned through 90 degrees suddenly becomes a letter 'd': we understand it is still an ampersand but has taken on a further meaning.

There is a further connotation of narrative that is also relevant to design. It is no surprise that many of the traditional forms of narration, whether epics or folk tales, are about journeys. Indeed a journey is often the starting point of a story, whether it is Odysseus seeking Penelope or Mr Mole abandoning spring-cleaning! The design process should also be a journey, not a process such as tinning sardines which has temporality but no deeper meaning. The design journey is an exploration and a discovery, both for designer and client. It is not the application of a style, nor the imposition of an inspired idea, but the disciplined understanding of the brief combined with the imaginative input of the designer's experience. There should be both structure and surprise in the outcome. Shouldn't all stories have unexpected endings?

character

Is this a fit time, said my father to himself, to talk of Pensions and Grenadiers?

Laurence Sterne: *Tristram Shandy*, 1759–1767

That one sentence is the whole of chapter five of book four of Laurence Sterne's *Tristram Shandy*. This novel was published in nine volumes between 1759 and 1767. Its appearance made the author's reputation, for its humour, its absurdity and its extravagance. Supposedly the autobiographical 'life and opinions of Tristram Shandy', the subject is not born until book three, and the main characters are Tristram's father, a man deluded by too much obscure learning, and his uncle, Toby, a retired officer with a mania for fortifications. Room enough for wit there, but above all its defiance of the conventions of narrative is what made, and makes, the book remarkable. The author/narrator (who is at times Sterne and at times Tristram) continuously breaks off his narrative to address the reader directly, to discuss how the book is going and how it should be written (should there be one chapter on chamber-maids and window-sashes or should each subject get one to itself, for a typical example). He plays endless games with the timing and structure of the story, propelling the reader in all directions, citing improbable authors (Hafen Slawkenbergius on noses is a favourite), and making an inordinate number of often bawdy jokes.

What is astonishing is that from these literary fireworks the characters emerge whole and rounded, be it the main figures or lesser ones such as the maid Suzannah or the modest parson Yorick. Their personalities are clear and coherent, their identities distinct.

The original concept of corporate identity was simply an identifying mark: this concept became subsumed into the idea of branding, a much more complex notion involving the expression of a company's ethos, beliefs and purposes, a statement of its mission. But on the way another concept got lost, that of the corporate character or personality. Not a personality in the sense of a visual cliché, like Michelin's Bibendum or the Pilsbury dough boy, but a sense of roundedness or completion achieved by the totality of the way the company presents itself to the outside world, whether to customers, stakeholders or the general public. This is not something that can be created by a single intervention, any more than we can get to know a person fully at a single meeting. It is a process that takes time and co-operation, and works best through a series of different projects rather than a single one, however long drawn-out, as this gives both client and designer the opportunity to look at the subject from a range of different perspectives.

There is often an assumption that a corporate identity has to be perceived as the same from all points of view, and this is generally true – identities get into difficulties when the perceptions do not match, as was the case for British Airways "Fly the World" campaign, brave effort though it was. But consistency does not have to be bland, and the way in which an identity is delivered does not have to be repetitive. Just as individuals have different facets to a personality and so are perceived as whole in different ways, it should be possible for a corporate identity to be delivered in different ways, through surprise and humour, for example, as much as conviction and integrity.

Bowers & Wilkins **"I don't really care how it's being recorded or what techniques and non-techniques are being used, I wanna hear the sound of how it feels." Dave Stewart is a songwriter, producer and performer who cares passionately about sound. He uses Bowers & Wilkins loudspeakers.**

The company was started by John Bowers and Peter Hayward in 1966 with a single, simple, passionate aim, to build the best loudspeakers possible. Through the changes in recording and delivering sound over the last forty years that aim has been maintained: B&W speakers are the equipment of choice in recording studios (including the famous Abbey Road studios in London) and in music-lovers' homes all over the world. Thomas Manss has been designing for B&W since 1995. The first commissions were graphic, for brochures and reports, but their involvement has now broadened considerably and now includes creating a promotional DVD, designing a trade fair stand and art directing B&W's website. "I sometimes think I know the company as well as some of its employees," Thomas comments. "Simply because we have been working together for so long." This involvement does give the Manss team a profound insight into the company, its aims and values. "I think of myself as a distance runner, not a sprinter," Thomas points out, "so this kind of enduring liaison suits me well."

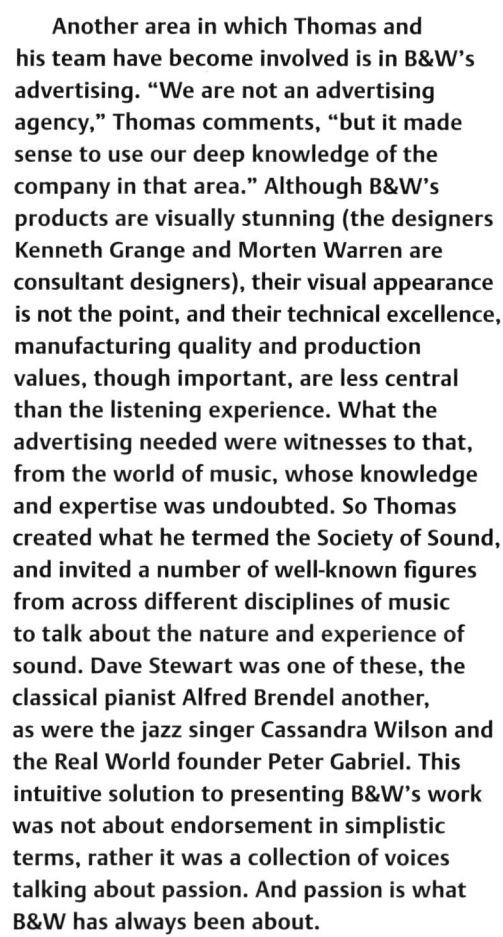

The changes Thomas has seen in B&W reflect the changes not just in technology but in the appreciation of sound. If in the 1970s hi-fi enthusiasts were the geeks of their era, today there is a much wider public appreciation of sound quality (one only has to compare an iPod with an early Walkman), and so B&W's market has expanded accordingly. The company has risen from being a cottage industry to being a major creator of acoustic systems. The trade fair stand Thomas is currently designing will not only feature B&W's classic Nautilus speaker, but the sound system B&W have designed for a new range of Jaguar cars as well as an iPod dock.

Another area in which Thomas and his team have become involved is in B&W's advertising. "We are not an advertising agency," Thomas comments, "but it made sense to use our deep knowledge of the company in that area." Although B&W's products are visually stunning (the designers Kenneth Grange and Morten Warren are consultant designers), their visual appearance is not the point, and their technical excellence, manufacturing quality and production values, though important, are less central than the listening experience. What the advertising needed were witnesses to that, from the world of music, whose knowledge and expertise was undoubted. So Thomas created what he termed the Society of Sound, and invited a number of well-known figures from across different disciplines of music to talk about the nature and experience of sound. Dave Stewart was one of these, the classical pianist Alfred Brendel another, as were the jazz singer Cassandra Wilson and the Real World founder Peter Gabriel. This intuitive solution to presenting B&W's work was not about endorsement in simplistic terms, rather it was a collection of voices talking about passion. And passion is what B&W has always been about.

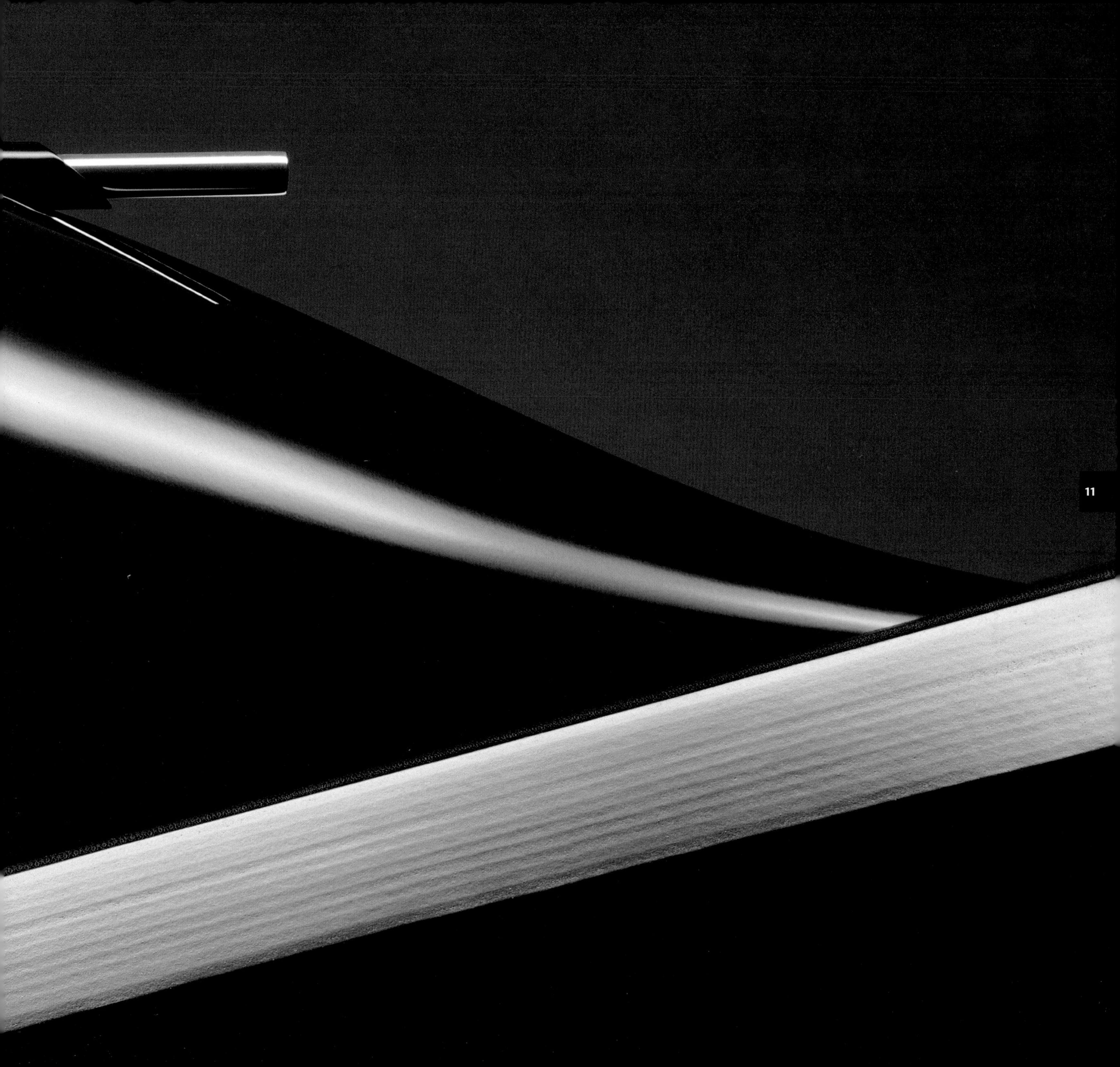

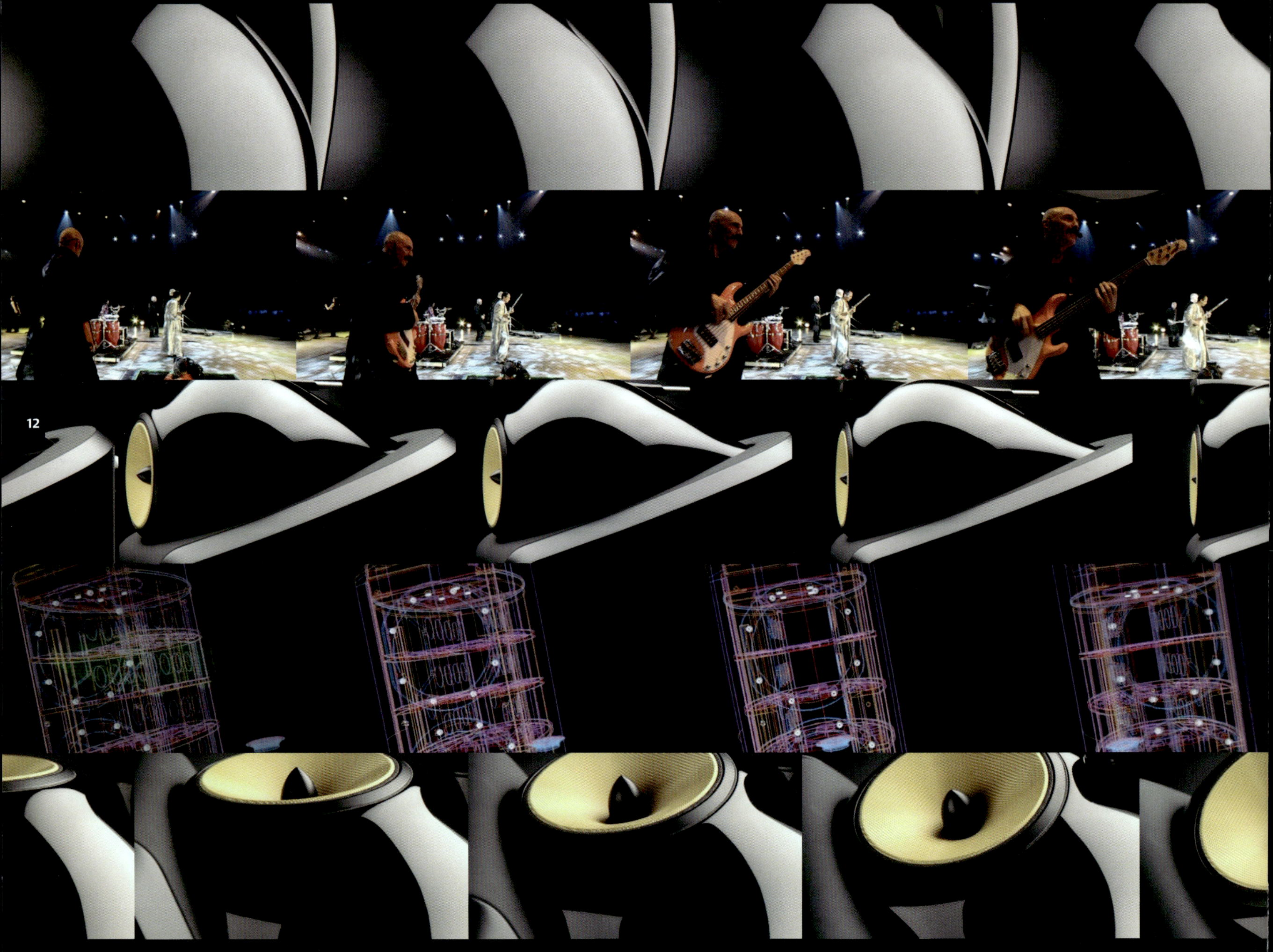

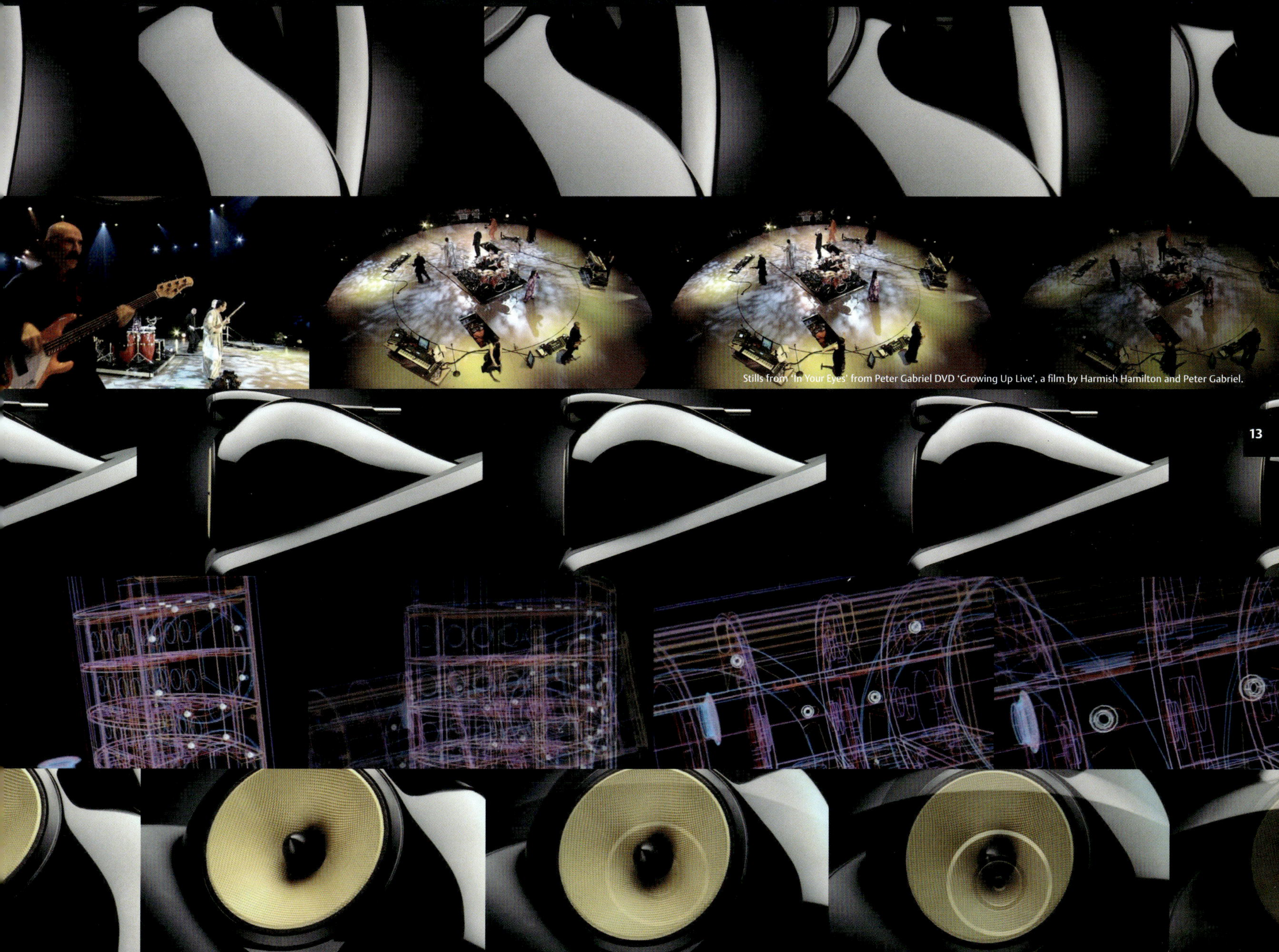

Stills from 'In Your Eyes' from Peter Gabriel DVD 'Growing Up Live', a film by Harmish Hamilton and Peter Gabriel.

B&W Bowers & Wilkins
800 Series
A Sound Experience
DVD
PAL

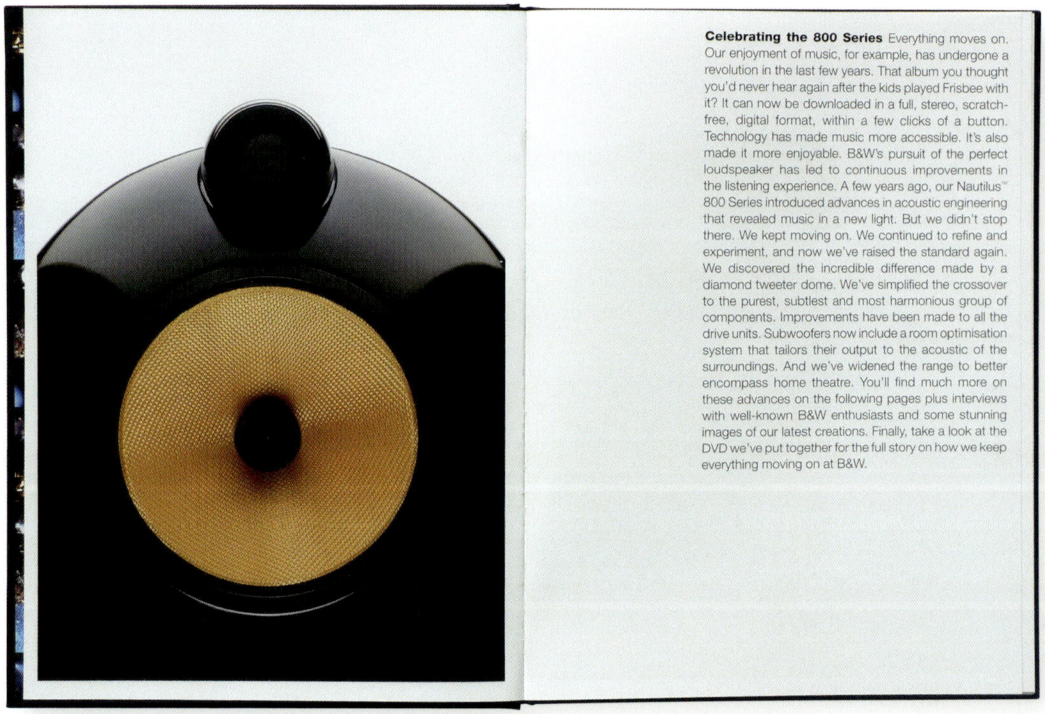
Celebrating the 800 Series Everything moves on. Our enjoyment of music, for example, has undergone a revolution in the last few years. That album you thought you'd never hear again after the kids played Frisbee with it? It can now be downloaded in a full, stereo, scratch-free, digital format, within a few clicks of a button. Technology has made music more accessible. It's also made it more enjoyable. B&W's pursuit of the perfect loudspeaker has led to continuous improvements in the listening experience. A few years ago, our Nautilus™ 800 Series introduced advances in acoustic engineering that revealed music in a new light. But we didn't stop there. We kept moving on. We continued to refine and experiment, and now we've raised the standard again. We discovered the incredible difference made by a diamond tweeter dome. We've simplified the crossover to the purest, subtlest and most harmonious group of components. Improvements have been made to all the drive units. Subwoofers now include a room optimisation system that tailors their output to the acoustic of the surroundings. And we've widened the range to better encompass home theatre. You'll find much more on these advances on the following pages plus interviews with well-known B&W enthusiasts and some stunning images of our latest creations. Finally, take a look at the DVD we've put together for the full story on how we keep everything moving on at B&W.

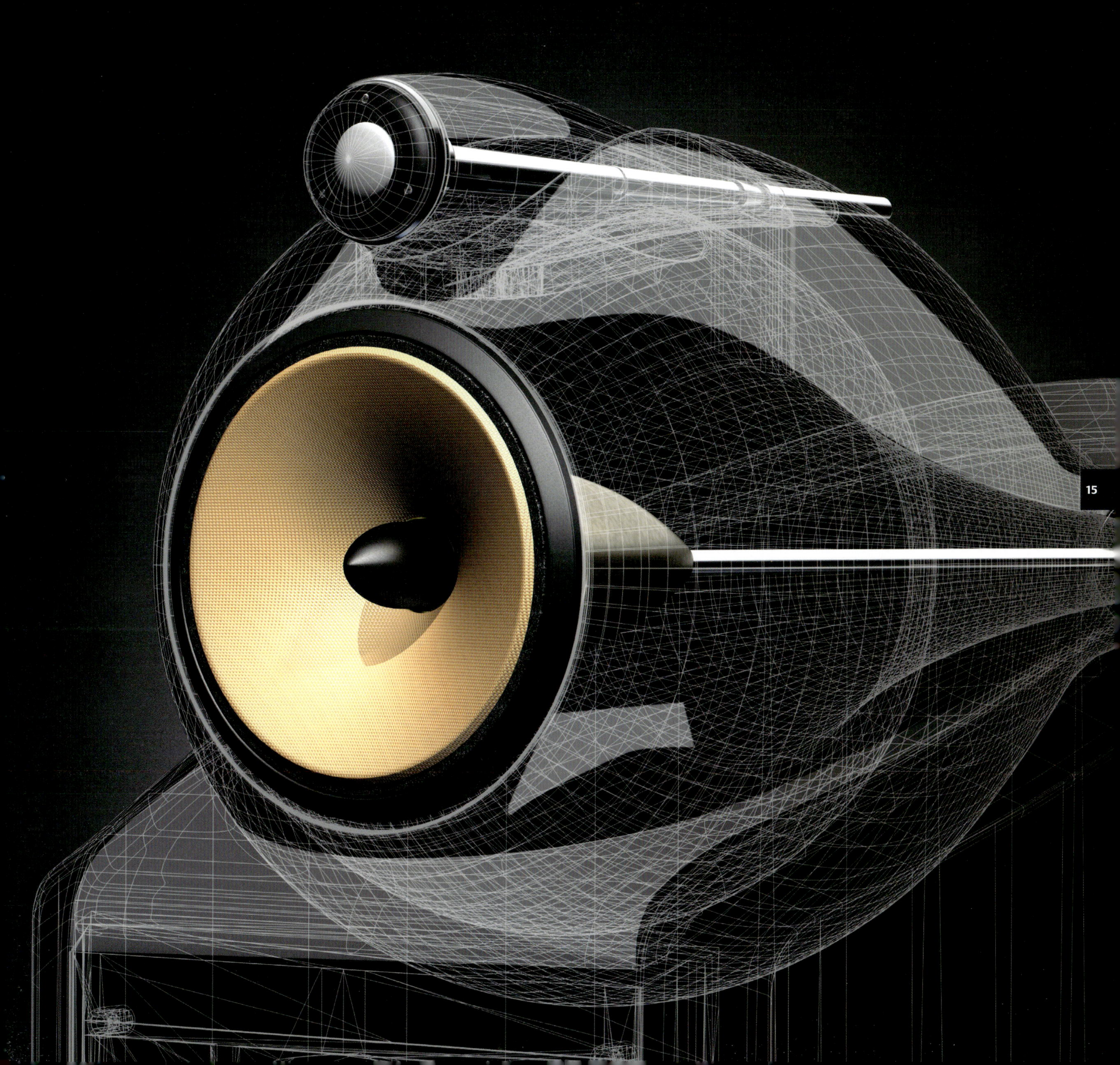

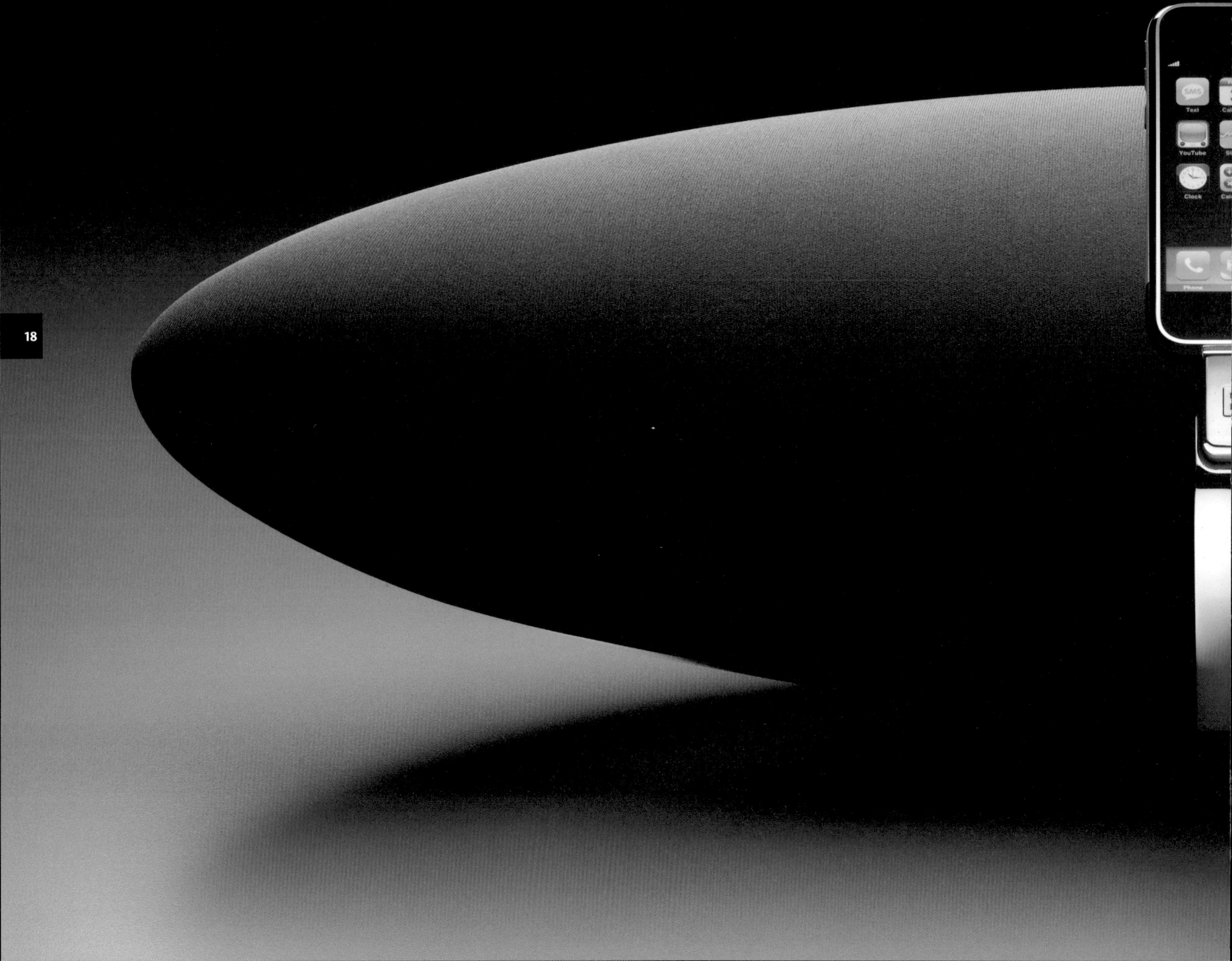
SMS
Text
YouTube
Clock
Phone

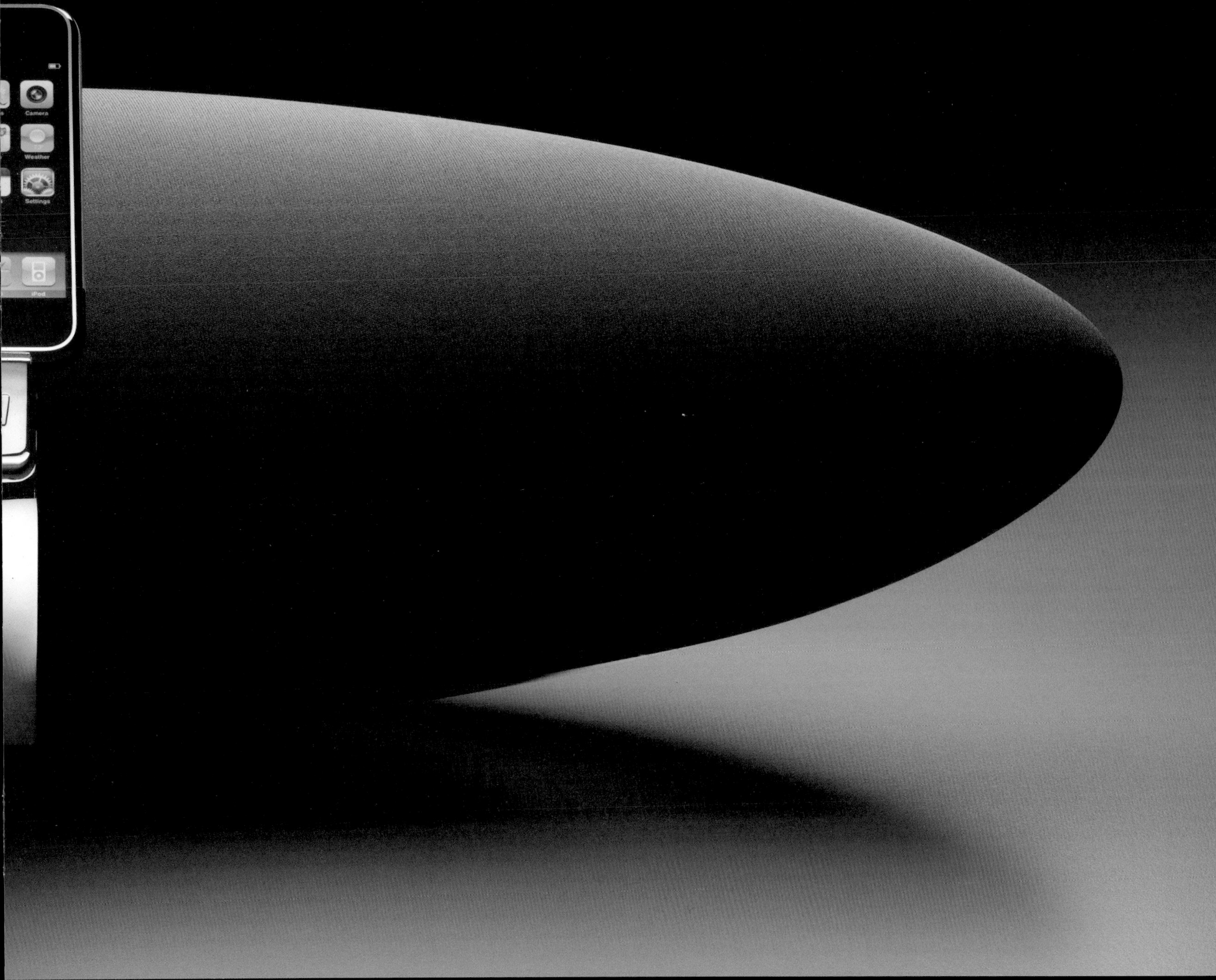
Camera
Weather
Settings
iPod

B&W
Bowers&Wilkins

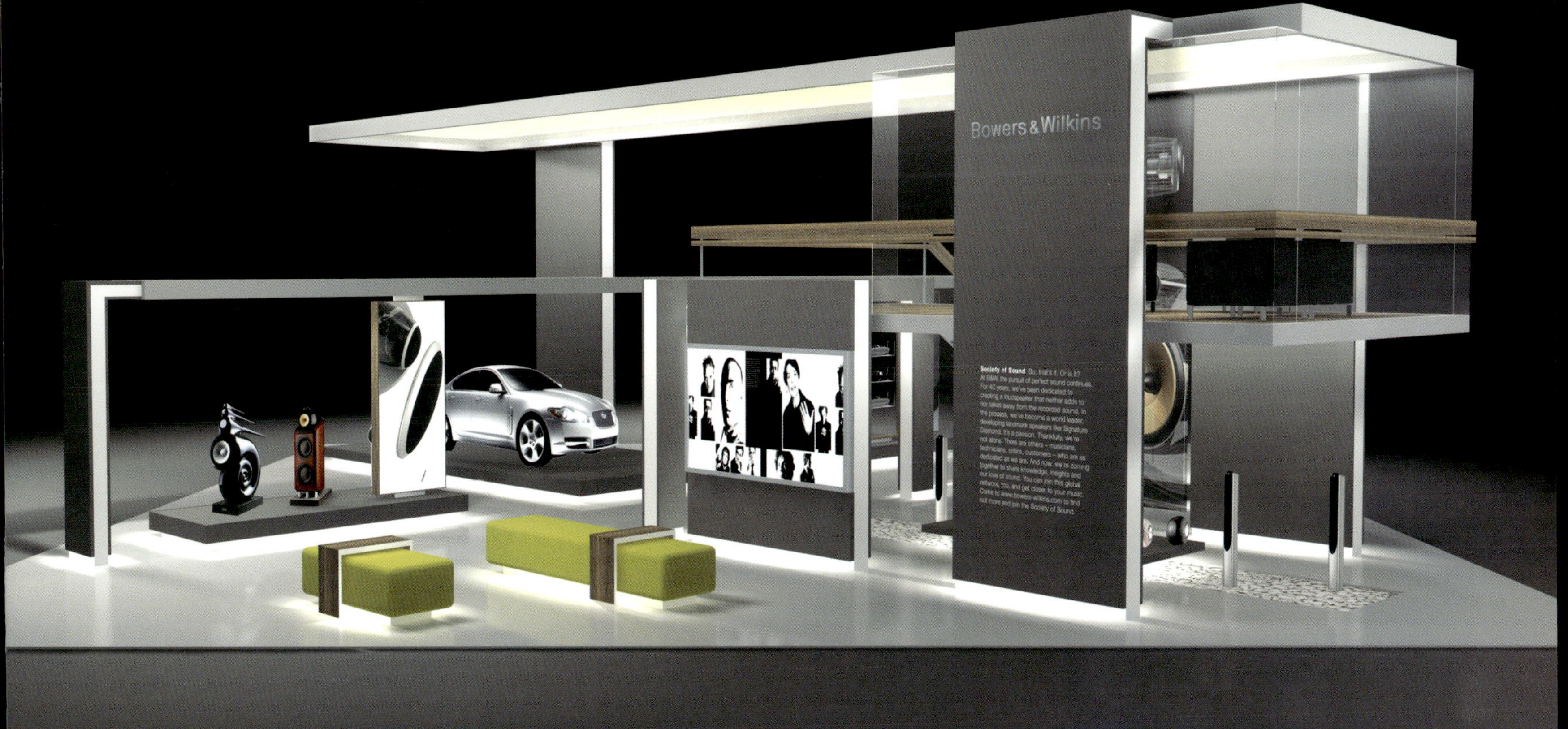
Bowers&Wilkins
Society of Sound

The Society of Sound is a meeting place for people who love sound. It's somewhere to come for inspiration and to find out about people who are using sound creatively. It's also a resource for hearing from people who have challenging ideas about the future of sound. Last but not least, the Society of Sound is an expression of Bowers & Wilkins' ongoing dedication to revealing sound in all its glory.

Society of Sound

passion

There is no sense in forgetting and every sense in dreaming. Thus the present is made rich. Thus the present is made whole, on the lagoon this morning, with past at my elbow, rowing beside me, I see the future glittering on the water. I catch sight of myself in the water and see in the distortions of my face what I might become. If I find her, how will my future be? I will find her. Somewhere between fear and sex passion is. Passion is not so much an emotion as a destiny.

Jeanette Winterson: *The Passion*, 1987

Jeanette Winterson's *The Passion*, her second novel after *Oranges are the Only Fruit*, traces the intersections in the lives of two characters which conventional analysis of the historical novel would describe as hero and heroine, though, as this is Winterson, it's the wrong analysis. One is Napoleon Bonaparte's cook, specifically the cook charged with ensuring that there was a roast chicken available to the Emperor at any time, night or day. The other is the daughter of a Venetian fisherman: tradition accords to such daughters the special quality of webbed toes. The trajectories of these historically minimal and slightly improbable characters traverse the period of the Napoleonic wars and their aftermath, but this is not a historical novel that seeks to causate history (in the way Neil Sullivan's *Newton trilogies* attempt) but rather an ironic, entertaining and elegant reminder of the primacy of the book's title, passion. These improbabilities (webbed toes and poulet rôtí) only serve to individuate the characters and immerse them in the wider theme: they are handled with Winterson's fluid and graceful prose so that they become completely credible to the reader. They are narrative devices that maintain the real metalanguage of the book (and of many of Winterson's other books) concerning the ineffabilities of the nature of passion and love (themselves distinguishable elements).

There are in any design project as well a series of trajectories and narratives. The most notable are those of the client's experience (or expectations) and the design agency's experience (or intentions). But there are further sets and subsets of experience, themselves trajectories or journeys, be it the individual user's experience of the design outcome or the learning curve of an individual designer or consultant involved in the project. The totality of these trajectories is the true history of the project (often impossible to reconstruct in detail) and so the final evaluation of it, the summation of the success or failure of the design seen as an event beginning with the brief and ending with the retirement of the design or its replacement with an alternative.

The designer cannot amass these trajectories and bring them to a conclusion or at least a closure in the way a writer of fiction can, but the designer can emulate the writer in anticipating potential trajectories. How will that or this group of users engage with this or that solution, what will their experience, their journey through or with it, be like? Will it match their demands and expectations or not?

Part of this information can be assessed though techniques such as content analysis, service design processes and market research, wherever each is appropriate, but there is also an overriding concept that will help designers manage this seemingly impossible task. That is, in a word, again, passion.

Oberhavel Verkehrsgesellschaft Oberhavel is a region of the former German Democratic Republic – East Germany – sitting to the north of Berlin. After reunification, the local utilities companies that had formerly been financed by and reported to state and local government found themselves in a new economic environment, subject to the rules of capitalism and competition, even though subsidies were available under the new regime (as they were in the rest of West Germany) for a large number of utility functions. So three local utility enterprises, newly privatised, decided to form themselves into a group under a holding company which would act as negotiator with government agencies. These were the local bus company (OVG), the refuse collection and recycling operation (GfA), and an agency for inward investment in the area (WfO). As such, they would of course need an integrated corporate identity programme that would unite them visually and frame their co-ordinated activities for government, stakeholders, their employees and clients as well as the general public. That is how corporate identity is defined and implemented in the West, of which they were now a part.

That is one trajectory: there is another. OVG had ordered eighty new buses from Mercedes, and one day Mercedes rang up and asked the fairly obvious question "what colour do you want them?" The managers of OVG went to see Thomas Manss, then visiting professor for corporate identity at the university in Potsdam, and asked him how to reply. He pointed out that any solution for OVG would have to work in parallel across the two other subsidiary companies, and upwards to the holding company. And along the sides of a bus, as well. But a solution was possible.

Here the two trajectories coalesce into one history, in that at the end of the day the Oberhavel group had a linked, articulated corporate identity. It was driven (no pun intended) by the bus group, in part since their need was the most urgent, and in part because the buses were the most visible component of the consortium. "Buses carry passengers," Thomas points out, "but they also carry advertising. Use a complex design, and once the advertising is added it looks like a dog's dinner: use a simple design and if there is no advertising the result is bland." The basic unit of the bus design was two pairs of low contrast colours: one was a pair of blues, medium and dark. This pair in turn formed the basis of the other identities as well. The second pair was a brighter blue and a pale green: together these four colours were used on the uniforms of the staff, on promotional items and in literature.

On the bus itself, the main body was decorated with horizontal stripes of the two blues, which would work well with or without advertising. "Then I found out that there is one part of a bus where there is never any advertising allowed, and that is around the front and side of the driver's seat. So I added stripes of the two secondary colours, which completed the design." Mercedes were not too happy with this, pointing out the additional spraying work that would need to be done if a panel was dented or damaged. "So I went to see the maintenance team at OVG," Thomas Manss recalls, "and they said they didn't bother with sprays: masking tape and a paint roller were far quicker." So the design stayed.

OVG
CITARO
OHV VK 90

The other identities also used the two blues in horizontal stripes, intercut with a second colour, green for the recycling company, red for the investment company, in the latter case rising like a graph of prosperity. The light blue, green and red are incorporated diagonally into the logo for the holding company. The initials of each company appear in white in the top corner of each logo.

"I would very much like to do more transport work," Thomas reflects, "for example an airline, because it is so complex and challenging. The bus project was an exciting one to do, particularly in bringing the client along the journey." But then Thomas Manss' design is always inflected with passion. And this passion can be shared by the clients: revisiting Oberhavel to photograph the work for this book, Thomas found the ten-year-old identity still as fresh as before, and strongly supported by management and staff. One driver had even bought a model bus and painted it by hand in the corporate colours: an idea that led to the company ordering a production run of models to be sold as souvenirs.

OVG
CITARO
OHV VK 80
Kleindiens

824 Oranienburg
OVG
OVG

OVG
OVG
OVG

OVG
CITARO
OHV VK 90

OH

OVG

WfO

GfA

desire

By a singular logic, the amorous subject perceives the other as a Whole … and, at the same time, this whole seems to him to involve a remainder, which he cannot express. It is this other as a whole who produces in him an aesthetic vision.

Roland Barthes: *A Lover's Discourse*, 1977

This series of fragments (the author's own term) are reflections of the feeling of a lover when separated from the object of love. The uncertainties and even despair of such a situation are often reflected in the texts, as are the moments of joy and hope. The references used are often literary: Goethe, Balzac, Proust and Verlaine, among many others, all provide models for consideration. Not surprisingly, many come from the Romantic canon, but this is more a disquisition on love than a celebration. If it seems at times elegiac, that is because it is endlessly serious. The frivolities of affection are not equated here with the hard work of loving another.

The book is also an investigation into the nature of identity, the meaning of knowledge and the role of language. To what extent can we know someone we love, or can we know even ourselves, are questions posed in various ways, and these link to Barthes' earlier interest in semiotics, myth and authorship. In his 1967 essay *The Death of the Author* he argues against the practice of readers using their knowledge of the author's tastes, opinions or context in understanding the meaning of a work: he renames the author as the scriptor, to cut the link between 'author' and 'authority'. Here he seems to be depersonalising the individuality of the lover, and of the loved one, as well as analysing whether a statement about a personal emotion can have any real meaningingfulness.

Barthes' insights have relevance to the design sector (we should not forget that his book *Mythologies* galvanised advertising). There are two aspects to this: firstly, the personality of the designer should not be part of the finished design: the design should seem to be appropriate and inevitable, the result of a natural process not a personalised intervention. The second aspect is desire: every design should be an object of desire, though that is not the same as a desirable object. Every design should be subtended by desire, in the serious sense of total commitment that Barthes so carefully analyses.

BITTE NICHT
STÖREN
PLEASE DO NOT
DISTURB
PRIÈRE DE
NE PAS
DÉRANGER

Meoclinic The Meoclinic in Berlin is a private luxury hospital. It offers top class medical services both for cosmetic work and medical needs in an atmosphere of comfort and well-being, to a diverse, wealthy, international clientèle. It is housed in a building on Friedrichstraße (Berlin's equivalent of Harley Street in London) designed by the architect I.M. Pei. The owners consulted Thomas Manss about the identity and interior design. The agency came up with two proposals. One was formal and very modern, the other more traditional. The client's immediate reaction was to go for the modern interpretation. But then Thomas Manss put a question to them: "Which setting and feel do you think your customers will prefer? They are the ones paying for the service." This was the crucial question: the clients would expect more traditional comfort, and so this was the solution adopted. Furniture by Eileen Gray, natural wood detailing, pale colours, all create the feeling of a hotel rather than a hospital.

For the logo Thomas Manss chose the phoenix as an emblem. A symbol of hope and regeneration, the mythical bird is known in classical Egyptian and later Arab writing, as well as in the Chinese tradition, where the Feng Huang is a symbol of grace and virtue. Stravinsky's opera, the Firebird, attests to the Russian version of the phoenix myth. Such a universally recognised symbol was appropriate for an institution catering to an international market. The logo is an ornate and graceful piece of calligraphy, and is used on signage as well as on cutlery, tableware and glassware. These are not the sturdy utilitarian products one might expect in a clinic, but of a standard and design to be expected in a fine restaurant. The logics of luxury and comfort are expressed so as to make a stay in hospital not only necessary but almost desirable.

The Meoclinic design is luxurious without being ostentatious. It is intended for wealthy and discerning patients, and seeks through the ambience it creates to meet their expectations. "Every piece of design," Thomas points out, "should be an object of desire." By this he means not merely desirable to the end user or client, or simply desirable in aesthetic terms for itself, but also created with desire on the part of the designer. A decade after its creation, and despite changes in the administration of the clinic and the services it offers, the branding of Meoclinic remains as robust and appropriate as ever.

MEOCLINIC

MEOCLINIC
Funktional
Anästhesiologie · Intensivmedizin
Schmerztherapie
Laboratoriumsmedizin
Pathologie
Physiotherapie · Rehabilitation
Krankengymnastik
Radiologie (MRT, CT, Sono)
Telemedizin
Gesundheits-Check-Up
Höhentraining · Sportmedizin
PST-Therapie

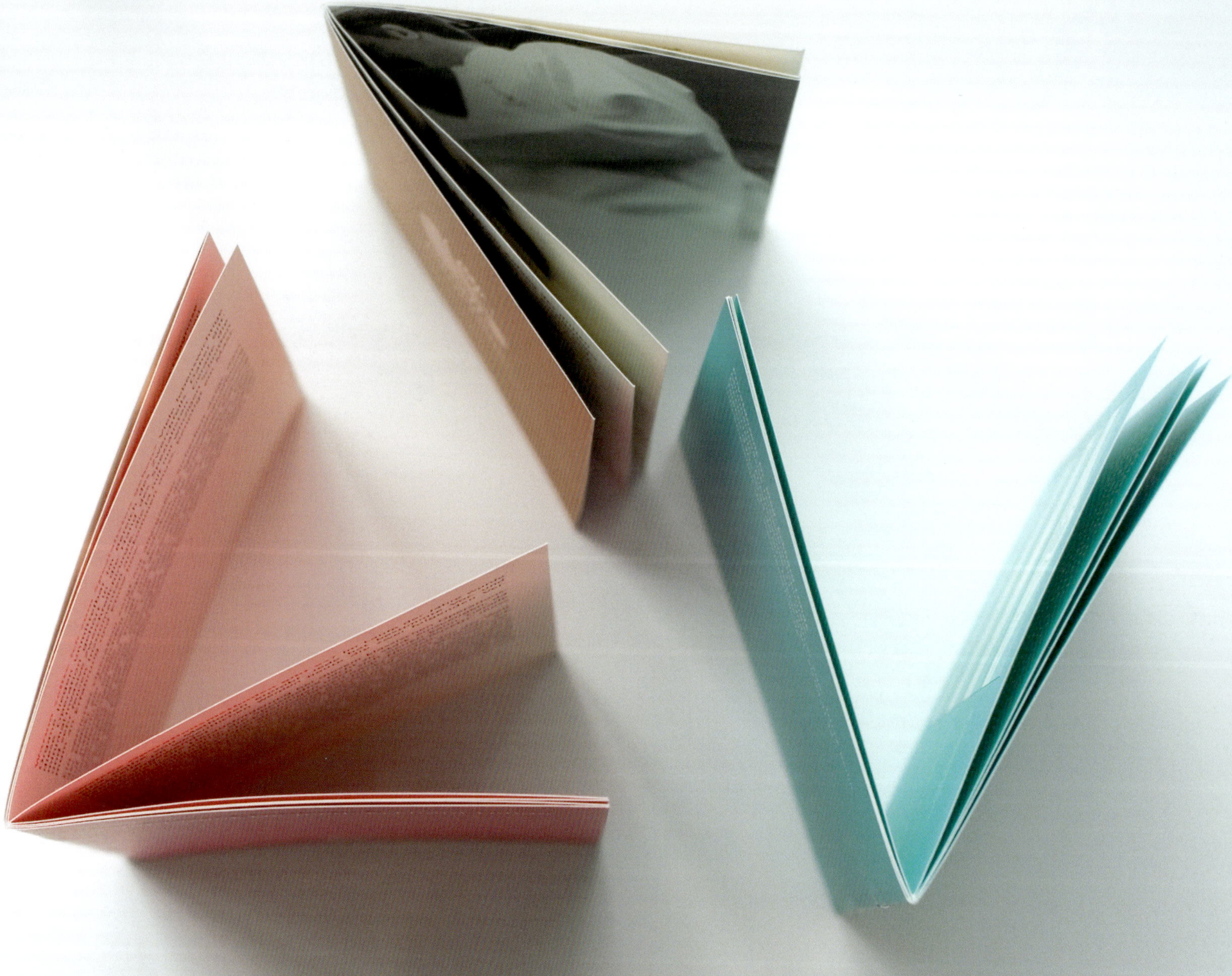

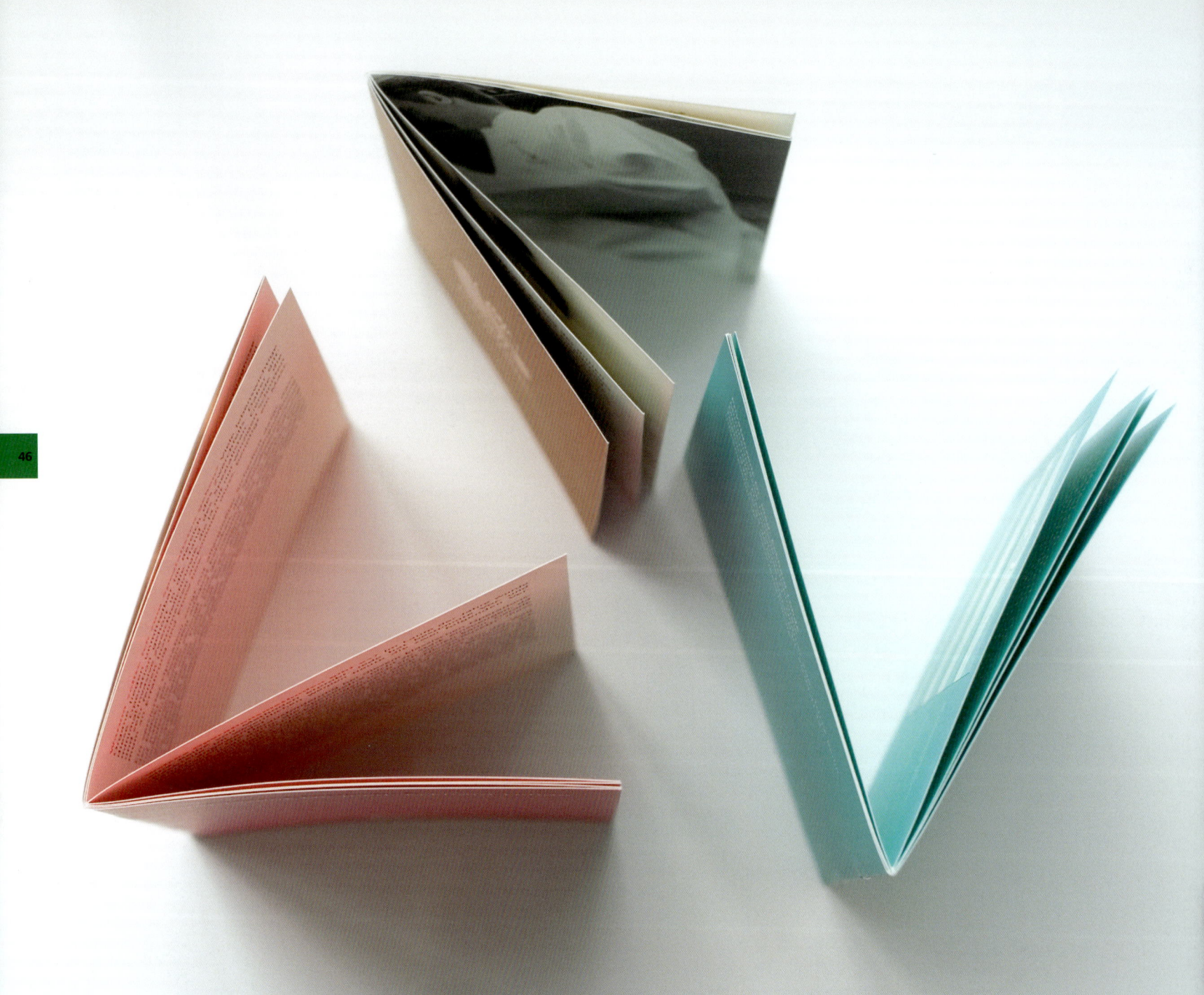

elegance

After the Widow left for the bazaar, the Scribe wavered at the door of his room, irresolute.The indentations of the reed mat where he had been lying for the past few days had left ribbed chain lines on the surface of his thighs. He ran his fingertips across the furrows of his skin and pondered his future. A scribe without paper was like a fish without water, he thought, half-tempted to write on himself.

Bahiyyih Nakhjavani: *Paper*, 2004

The exotic enters the novel in the nineteenth century, and in line with that century's view, at least from Europe, of the rest of the world, it does so in imperialism cloaked in the romantic, as in Verne or Kipling. One has to wait until the twentieth century for exotic cultures to speak in their own voices, and work within the definitions of their own culture, bringing the conventions of the European novel to a new range of subject matter and society from Africa, South America, India and Asia, and empowering a range of new voices.

This exquisitely structured novel by Bahiyyih Nakhjavani is set in the Persian frontier in the mid-nineteenth century. Subtitled *The Dreams of a Scribe,* it recounts the search for a perfect sheet of paper, for the writing of a masterpiece, at a time when traditional paper making is in decline under the pressure of imported European factory-made paper. The dreams guide the scribe on his search (perhaps) and through the characters of the Widow, the Mullah, the Thief, the Moneylender, the Envoy, the Governor and his Daughter, the author unveils a world of castes and customs under the pressure to change, unfamiliar and surprising, yet with themes of desire, temptation and perfection that we can identify and understand. The ending is unexpected but inevitable: intricate and satisfying as the pattern on a Persian rug.

The logics of successful surprise – the unexpected inevitability of a surprise – works in design as well. A logo that plays a visual trick, for example, works not because it is unexpected, but because the viewer can also sort out how it was done: it seems to have arisen naturally, almost. This is an elegant surprise: quite different from just shouting "boo". Achieving this too is a quest for perfection, demanding an understanding of the materials available (type, colour, surface, light and so on) and how they can seem to flow together effortlessly into the solution.

Oyuna The team behind Oyuna came to Thomas Manss for advice about their corporate identity and branding at the time the company was starting operations. The intention, they explained, was to produce their cashmere designs in Mongolia, and sell them initially through other retailers. It would begin as a business-to-business enterprise offering cashmere pieces for fashion and accessories, the home, and interior design. Once this was established, they would extend into direct retailing. Cashmere is a luxury product anyway: they wanted to be at the luxury end of the luxury market. As Thomas Manss put it, "they wanted to look at home amongst the exclusive boutiques in Knightsbridge, even though it might take a while until they moved in."

The company had originally thought of the name *Cashmere Republic* but felt that it did not have the right feel to it. Thomas agreed: the name was too trivial. The founder of the company is called Oyuna Tserendorj: how about using her first name as the corporate name? There were precedents enough in the world of fashion, after all. The name is a Mongolian one, so it is both exotic and apposite.

Cashmere is both classic and modern, so Thomas used a serif typeface, Modern No 216 light, setting the name in lower case. Then he cut the right branch off the letter 'y'.

At first glance the eye supplies the missing piece – the upper serif on the adjacent 'u' helps this, as the spacing is closed up. But a closer look reveals the surprising gap, and one at once knows how it was done: the reader shares the surprise because it is understood. The name was used on labels, for letterheads and in advertising.

The client also wanted a website. Here the issue was not to offer finished work or sale, and not even for inspection (as then it would be too easy to copy any designs). So only details are used, framed and modelled to convey the richness and luxury of the material. There is a lightness of touch about the presentation: the horizontal images appear within a black border, with minimal text. "It is one of the most visually rich websites we have ever done, for its size," Thomas comments, "and yet there isn't – and this is deliberate – a complete garment or accessory anywhere." The result is like a shop window: one in Knightsbridge, that is.

oyuna
100%
cashmere

oyuna

oyuna
c a s h m e r e

oyuna
oyuna
oyuna
oyuna
Colours 2007
Collection 2006

Lounge Wear
Easy-wear cashmere
for stylish comfort
Mandu knitted wrap (one size)
charcoal
beige

Soko knitted throw with ribbed edge 180x120cm (70x47")
• bedspread 220x180cm (86x70")
soft grey with
slate grey edge
stone brown with
charcoal edge
cream with
ivory edge
The essentials of cashmere living.
to soothe, to relax,
Timeless

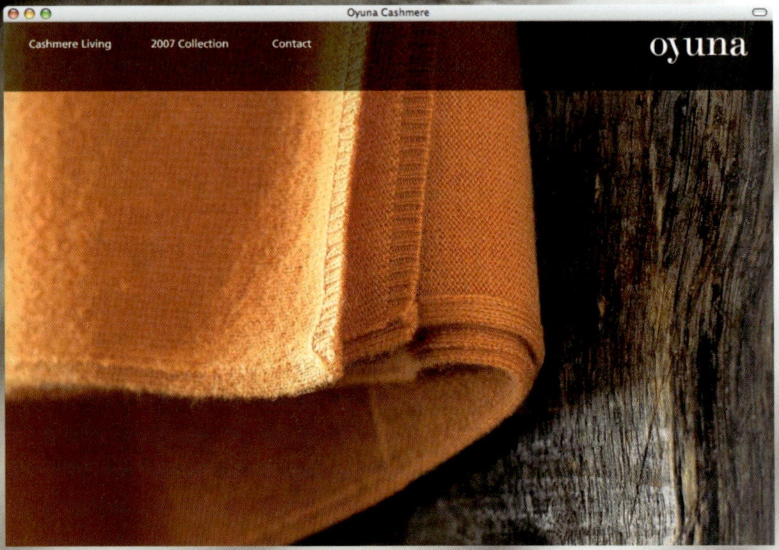

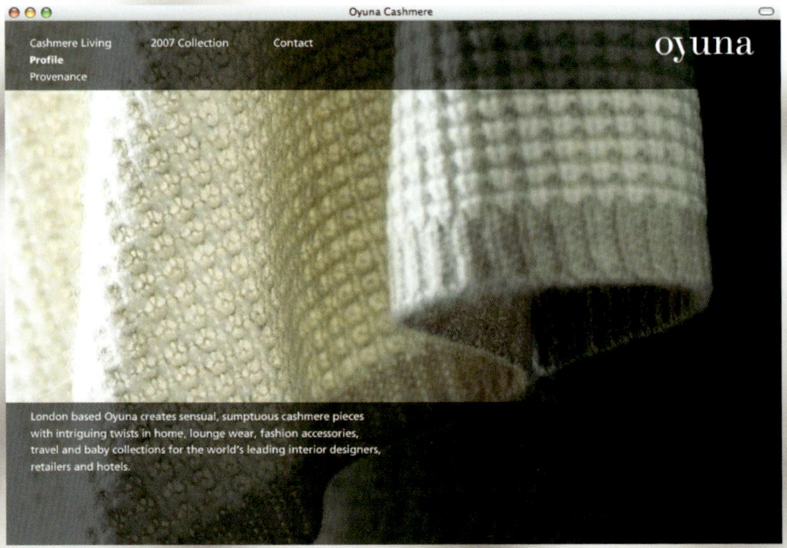

www.oyuna.com

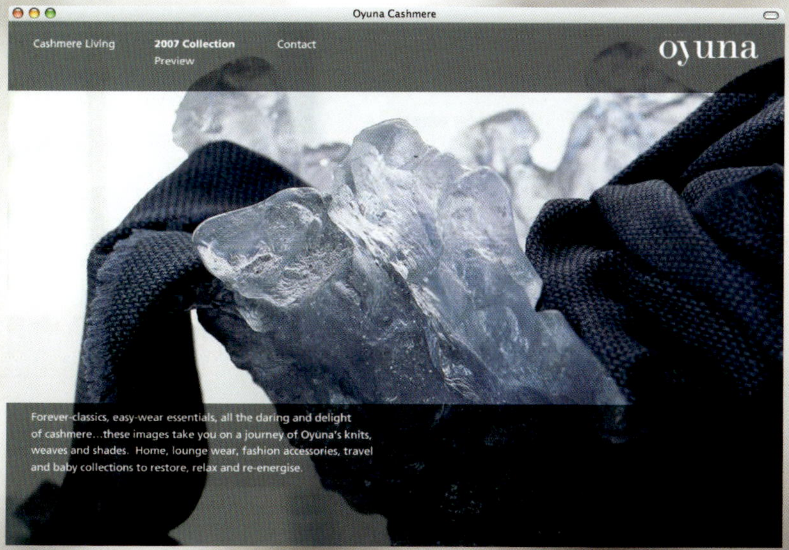
Oyuna Cashmere
Cashmere Living
2007 Collection
Preview
Contact
oyuna
Forever-classics, easy-wear essentials, all the daring and delight
of cashmere…these images take you on a journey of Oyuna's knits,
weaves and shades. Home, lounge wear, fashion accessories, travel
and baby collections to restore, relax and re-energise.

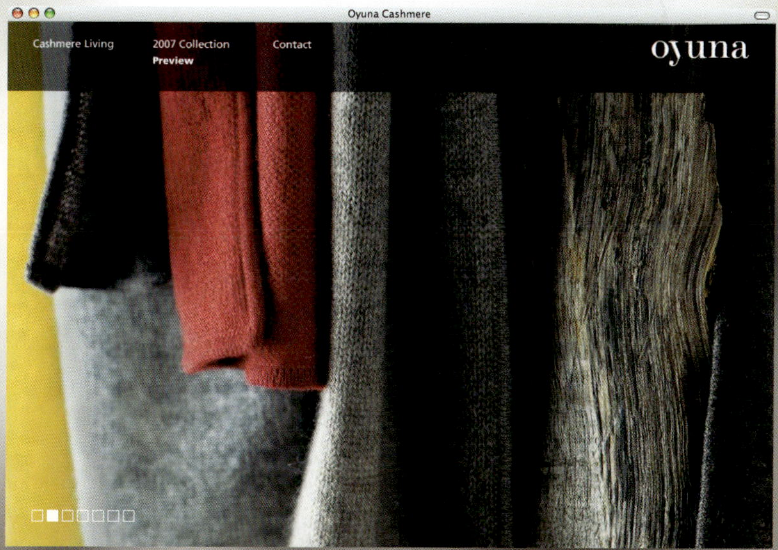
Oyuna Cashmere
Cashmere Living
2007 Collection
Preview
Contact
oyuna

continuity...

Here we will explore the idea that the different structure-generating processes that result in meshworks and hierarchies may also account for the systemacity that defines and distinguishes every language. In particular, each vowel and consonant, each semantic label and syntactic pattern, will be thought of as a replicator.

Manuel De Landa: *One Thousand Years of Non-Linear History*, 1997

Manuel De Landa's *One Thousand Years of Non-Linear History* contains no dates or battles, queens or emperors. It looks instead at the history of the last one thousand years as a set of geological, biological and linguistic histories. Dealing with themes such as urbanisation, trade and agriculture, the author, an artist and film-maker as well as an academic, seeks to understand not the specifics of history but its thematics. He looks at the way that systems interact to create changes, arguing that the logics that apply to geology are also applicable to biology and linguistics. This is the main argument of the book, that we need to understand systems, meshworks, hierarchies and replicators, to understand both history and the current state of the world. It is an argument that has repercussions in our understanding of evolution, economics and the nature of human thinking. De Landa's is a complex and engaging argument, and at first its relevance is unclear. The idea that there are similarities between the development of geological forms and of languages seems strange. But de Landa uses the idea of 'abstract machines' developed by Guattari and Deleuze and shows that such entities are necessary if one is to avoid vague notions such as polity or society in trying to account for change. Considering institutions not just in terms of their structure and regulation, but instead as distributors of energy (financial, political and so on) allows one to make these challenging comparisons.

But what it does stress is the importance of time, and of taking a broad view. Both of these are qualities that a designer needs. Being able to see 'beyond the brief' has long been acknowledged as the key to many successful designs, but equally important is the investment in time in getting to know and understand a client. It does take time to learn the client's language (and for the client to learn the designer's means of expression). This kind of continuous exchange creates a base of knowledge that informs the development of design work.

Foster + Partners Norman Foster opened his own architectural practice in 1967. Forty years on, and Foster + Partners have an international reputation for major building projects around the world. Often using complex but elegant engineering solutions, the practice's work has won many awards. Norman Foster himself was knighted in 1990 and made a life peer in 1999: in the same year he won the prestigious Pritzker Prize. There is a particular problem in writing about an architect's work, which is about how to date it. Does one date a design from its conception – the original commission or the winning of a competition, for example – or by the completion of the building of a project. If the former, how to account for any changes in the development of the project from conception to build, while if the latter date is taken, the history of the design development of the practice is falsified: such and such an idea does not date from 2005, say, when the building is finished, but from 2001 when the design was first conceived. In short, conventional processes of time and date do not apply. What is needed is an evolutionary or developmental line of access, in which conceptual values – such as the importance of environmental responsibility or the potential of engineering, to take two examples in Foster's case – can be traced across a meshwork of different projects, some built, some unbuilt, some for architecture, some for design, from which a coherent picture of the worth and importance of a body of work can emerge.

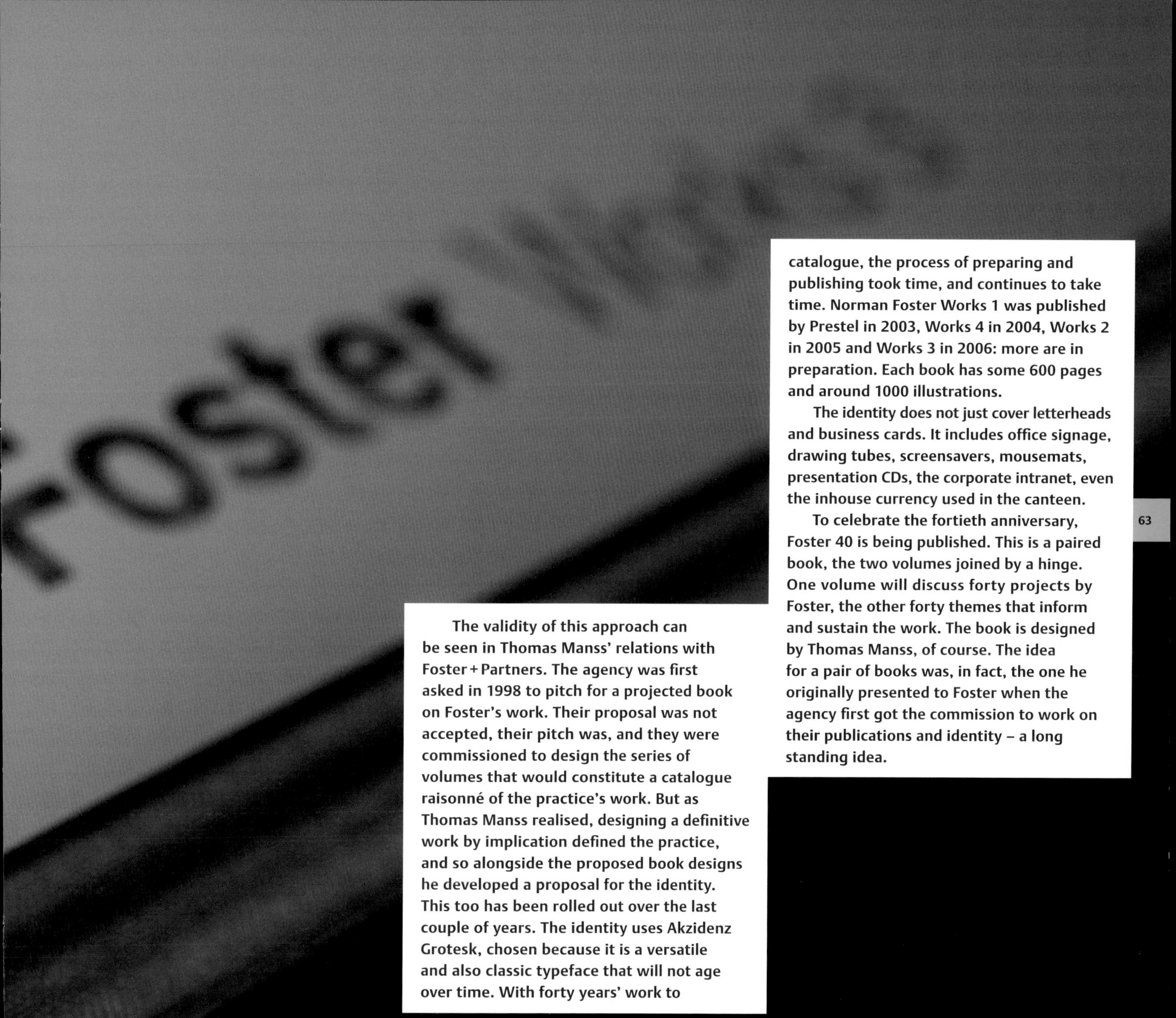

The validity of this approach can be seen in Thomas Manss' relations with Foster + Partners. The agency was first asked in 1998 to pitch for a projected book on Foster's work. Their proposal was not accepted, their pitch was, and they were commissioned to design the series of volumes that would constitute a catalogue raisonné of the practice's work. But as Thomas Manss realised, designing a definitive work by implication defined the practice, and so alongside the proposed book designs he developed a proposal for the identity. This too has been rolled out over the last couple of years. The identity uses Akzidenz Grotesk, chosen because it is a versatile and also classic typeface that will not age over time. With forty years' work to catalogue, the process of preparing and publishing took time, and continues to take time. Norman Foster Works 1 was published by Prestel in 2003, Works 4 in 2004, Works 2 in 2005 and Works 3 in 2006: more are in preparation. Each book has some 600 pages and around 1000 illustrations.

The identity does not just cover letterheads and business cards. It includes office signage, drawing tubes, screensavers, mousemats, presentation CDs, the corporate intranet, even the inhouse currency used in the canteen.

To celebrate the fortieth anniversary, Foster 40 is being published. This is a paired book, the two volumes joined by a hinge. One volume will discuss forty projects by Foster, the other forty themes that inform and sustain the work. The book is designed by Thomas Manss, of course. The idea for a pair of books was, in fact, the one he originally presented to Foster when the agency first got the commission to work on their publications and identity – a long standing idea.

Norman Foster Works 4

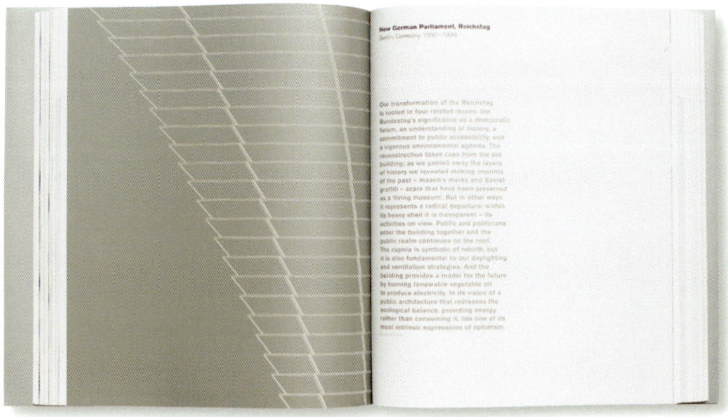
New German Parliament, Reichstag
Berlin, Germany 1992 – 1999
Our transformation of the Reichstag is rooted in four related issues: the Bundestag's significance as a democratic forum, an understanding of history, a commitment to public accessibility and a vigorous environmental agenda. The reconstruction takes cues from the old building: as we peeled away the layers of history we revealed striking imprints of the past – mason's marks and Soviet graffiti – scars that have been preserved as a 'living museum'. But in other ways it represents a radical departure: within its heavy shell it is transparent – its activities on view. Public and politicians enter the building together and the public realm continues on the roof. The cupola is symbolic of rebirth, but it is also fundamental to our daylighting and ventilation strategies. And the building provides a model for the future by burning renewable vegetable oil to produce electricity. In its vision of a public architecture that redresses the ecological balance, providing energy rather than consuming it, lies one of its most intrinsic expressions of optimism.

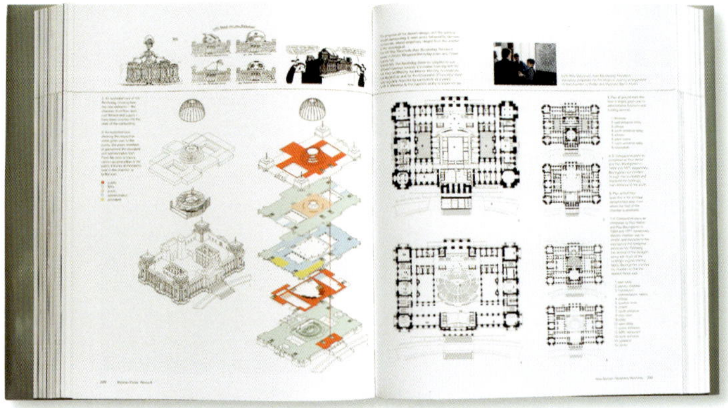

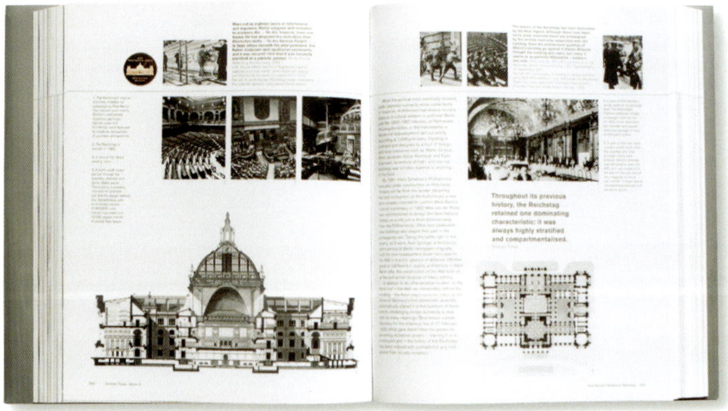
Throughout its previous history, the Reichstag retained one dominating characteristic: it was always highly stratified and compartmentalised.

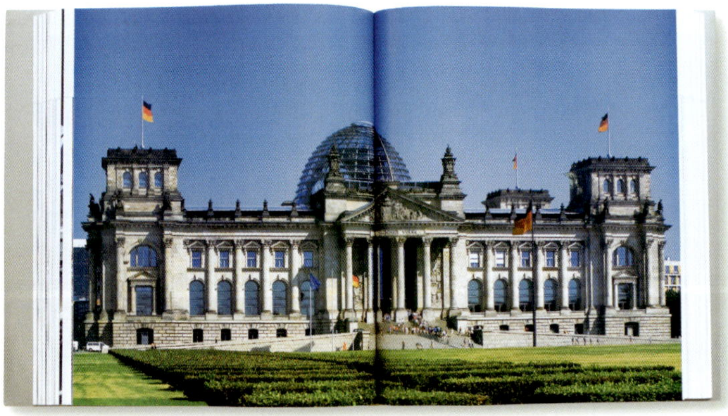

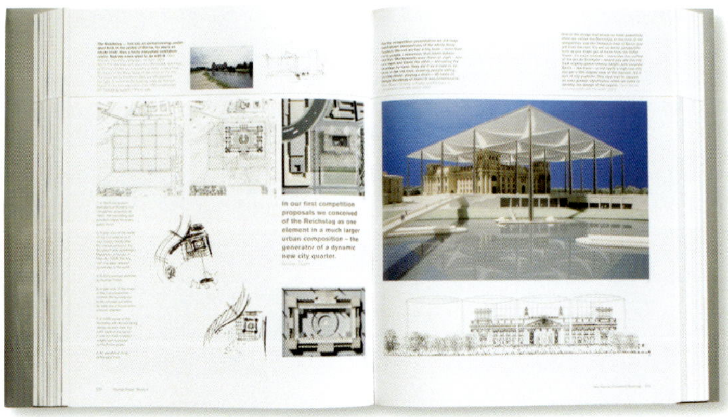
In our first competition proposals we conceived of the Reichstag as one element in a much larger urban composition – the generator of a dynamic new city quarter.

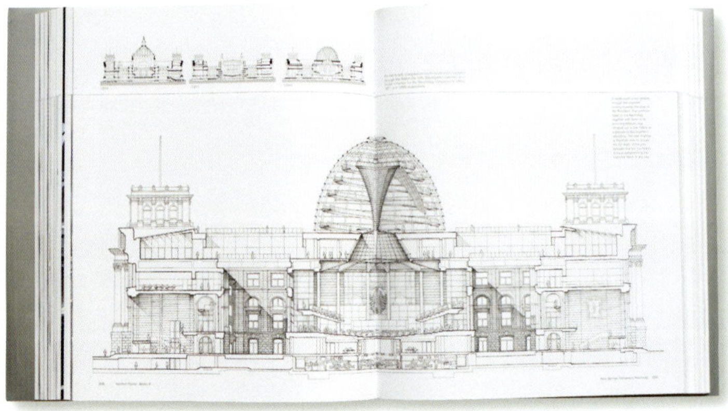

Sculptures and paintings [installed] throughout the Reichstag tell the narrative of twentieth-century history. Positioned carefully, sometimes even poignantly, these art works trigger a response that is sometimes joyful, sometimes spiritual. For many months Foster worked with specially commissioned German and Russian, French and American artists. Nonie Niesewand, The Independent, 19 April 1999
Left: While visiting Gerhard Richter's studio in Cologne in July 1998 Norman Foster studies a model of an early version of the artist's installation of the German flag in the Reichstag lobby.
Right: Norman Foster and Mark Braun discuss the project with Gerhard Richter (left).

Previous pages: The west front of the Reichstag photographed in the early afternoon. The line of visitors on the steps is part of the building's daily routine.

1. A view of the west entrance hall with its monumental German flag executed in glass by Gerhard Richter. It is from this space that visitors are taken by lift to either the mezzanine-level tribunes or the roof deck and the cupola with its panoramic viewing deck.

2. Facing Richter's flag in the west lobby is an installation by Sigmar Polke. The illuminated boxes have a viewing surface of prismatic plastic, so the images appear to move as the eye travels across them. The panels incorporate references to iconic German figures, from Bismarck to Konrad Adenauer.

3. A view along the west lobby in front of the chamber, looking north.

1

The reconstructed building learns from its history. We have reinstated the original entry sequence, up the grand flight of steps from the west, so that public and politicians enter the Reichstag as equals, by the same route.
Norman Foster

Sigmar Polke, the reigning court jester of German art, is able to give the new Reichstag the spin it deserves … His five holographic collages poke fun at key figures and symbols of German politics … These significantly dense works transform as you walk past them, like elaborate versions of a winking Jesus postcard. With antic irreverence they call the bluff of Foster's rhetorical exercise in transparency. Reality, they remind us, is never transparent. What we see from any one angle is always only part of the picture. Ralph Rugoff, The Financial Times, 19 April 1999

The 1960s [remodelling of the Reichstag] reflected the self-conscious modesty and insecurity of the young West German Republic. The exterior was stripped down and robbed of its dome. The utilitarian interior resembled a mediocre airport lounge. [illegible], The Financial Times, 17 April 1999
Right: The west lobby as completed by Paul Baumgarten in 1971 (compare this with the view below).

4

4. Looking into the chamber from the west entrance hall. Arriving visitors have a clear view of the seats of the Chancellor, the President of the Bundestag and other national leaders.

5. Visitors thronging on the Reichstag's steps, waiting to enter the building.

40 Projects
40 Themes

Foster 40 Themes

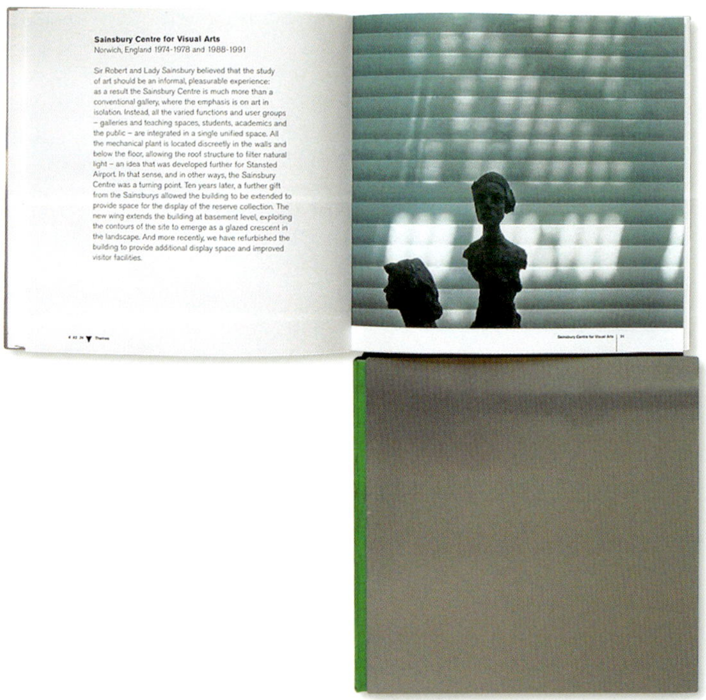
Sainsbury Centre for Visual Arts
Norwich, England 1974-1978 and 1988-1991
Sir Robert and Lady Sainsbury believed that the study of art should be an informal, pleasurable experience: as a result the Sainsbury Centre is much more than a conventional gallery, where the emphasis is on art in isolation. Instead, all the varied functions and user groups – galleries and teaching spaces, students, academics and the public – are integrated in a single unified space. All the mechanical plant is located discreetly in the walls and below the floor, allowing the roof structure to filter natural light – an idea that was developed further for Stansted Airport. In that sense, and in other ways, the Sainsbury Centre was a turning point. Ten years later, a further gift from the Sainsburys allowed the building to be extended to provide space for the display of the reserve collection. The new wing extends the building at basement level, exploiting the contours of the site to emerge as a glazed crescent in the landscape. And more recently, we have refurbished the building to provide additional display space and improved visitor facilities.
Themes

40 Projects
40 Themes

Foster 40

Canopies
While technology can help us to regulate a building's environment, there are also valuable lessons to be learned from regional traditions and centuries-old techniques of passive environmental control – for example, using oversailing roofs or fabric canopies to shelter us from the rain or deflect the heat of the sun.

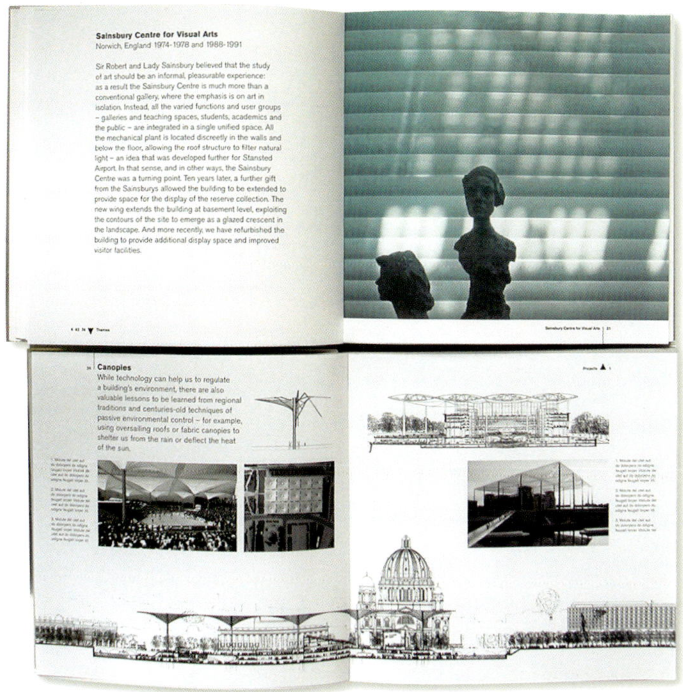
Sainsbury Centre for Visual Arts
Norwich, England 1974-1978 and 1988-1991
Sir Robert and Lady Sainsbury believed that the study of art should be an informal, pleasurable experience: as a result the Sainsbury Centre is much more than a conventional gallery, where the emphasis is on art in isolation. Instead, all the varied functions and user groups – galleries and teaching spaces, students, academics and the public – are integrated in a single unified space. All the mechanical plant is located discreetly in the walls and below the floor, allowing the roof structure to filter natural light – an idea that was developed further for Stansted Airport. In that sense, and in other ways, the Sainsbury Centre was a turning point. Ten years later, a further gift from the Sainsburys allowed the building to be extended to provide space for the display of the reserve collection. The new wing extends the building at basement level, exploiting the contours of the site to emerge as a glazed crescent in the landscape. And more recently, we have refurbished the building to provide additional display space and improved visitor facilities.
Themes
Canopies
While technology can help us to regulate a building's environment, there are also valuable lessons to be learned from regional traditions and centuries-old techniques of passive environmental control – for example, using oversailing roofs or fabric canopies to shelter us from the rain or deflect the heat of the sun.

Fost

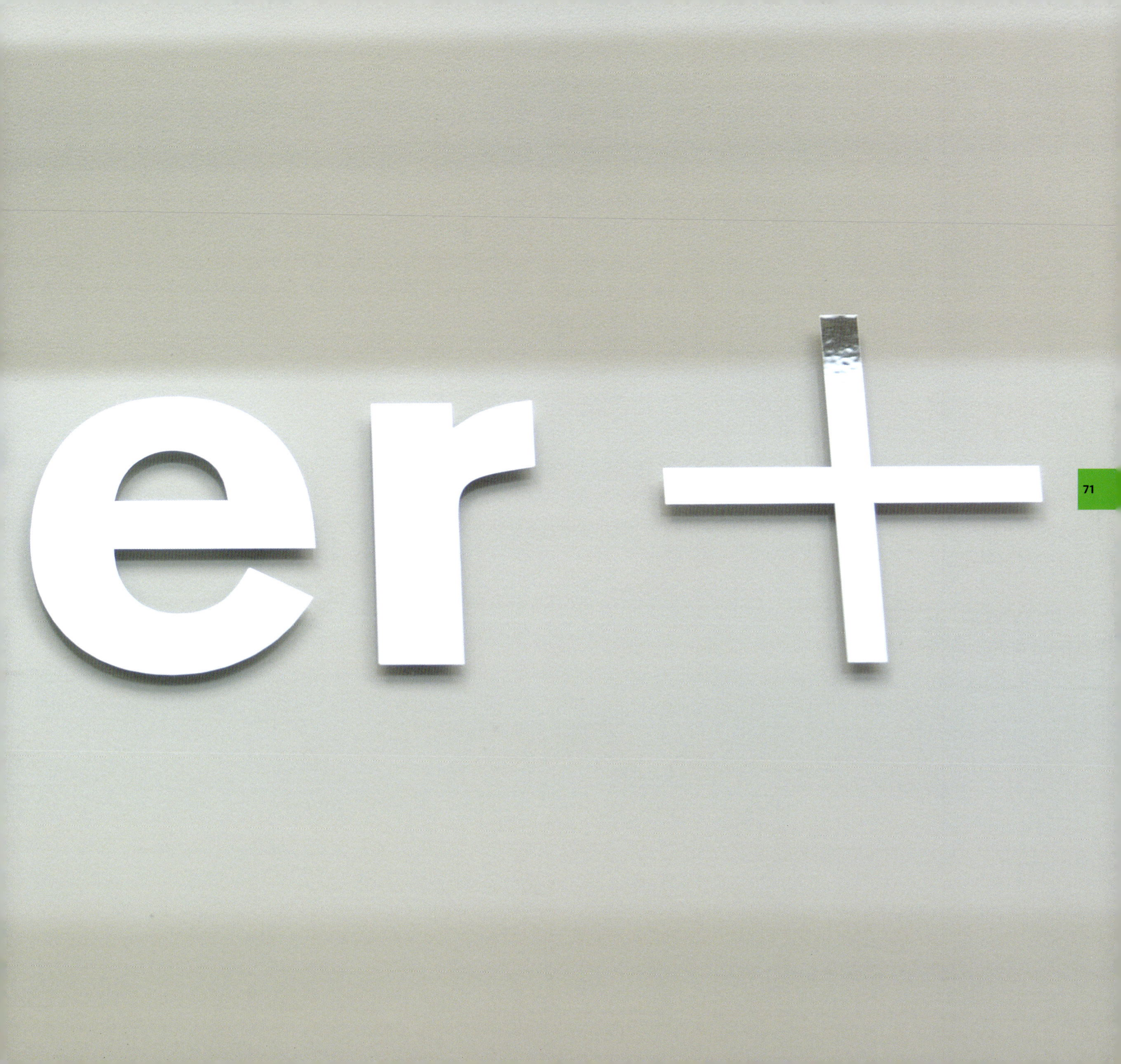
er +

Foster + Partners

Foster + Partners
Katy Harris BA (Hons)
Partner
Riverside, 22 Hester Road
London SW11 4AN
T +44 (0)20 7738 0455
D +44 (0)20 7943 6266
F +44 (0)20 7738 1107
kharris@fosterandpartners.com

Foster + Partners

Foster + Partners

Foster + Partners

www.fosterandpartners.com

Practice Projects Team News Data Contact
Foster + Partners
Practice Projects Team News Data Contact
Foster + Partners
Practice Projects Team News Data Contact
Foster + Partners
Practice Projects Team News Data Contact
Foster + Partners
Practice Projects Team News Data Contact
Foster + Partners

responsibility

Two monks were watching a flag flittering in the wind. One said, 'it's the wind that moves.' The other said, 'I disagree, it's the flag that moves.' But a Zen patriarch standing nearby, said, 'it's not the wind or the flag ... it's the mind that moves.'

Alan Fletcher: *The Art of Looking Sideways*, 2001

Alan Fletcher's career until his sad death in 2006 is in itself a history of the rise of graphic design in Britain in the twentieth century. Starting with the partnership Crosby Fletcher Forbes in the 1960s, he then became one of the founding partners of Pentagram, the partnership that woke corporate Britain up to the value of design. Retiring from Pentagram to start his own studio, he presided over the reinvention of the art book in his work for Phaidon. They published *The Art of Looking Sideways*, a book of his accumulated quotations, images, comments, diary entries and stories. The design is ever-changing, adapted in masterly fashion to the subject matter. The book appears to explore an endless series of different themes: one can open it at random and read a quotation or a story, often elegantly witty, or come across a description of a rare word. The images range from found objects, Alan's own sketches and designs, to visual puns and typographic devices. The book is the fruit of eighteen years of accumulating notions, images and ideas.

At first sight it seems formless, an elegant scrapbook. But as one gets beneath the surface effects, stimulating as they are, it is clear that Fletcher is making a multiple argument for the importance of understanding design and designing intelligently. The analysis of the nature of symbols – graphic, visual, typographic – is profound, as is the account of human perception and the relation of hand, eye and brain. His serious fascination with the symbolic is not only relevant to other designers and design users, but also illuminates his own design practice. Symbolic functions, whether through typography, the social use of colour, or other means, are at the heart of design practice and as such of key importance in finding our way around the modern, designed world.

COLLECTION 2005
NAUMAN

Tate Symbolism and clarity are of particular importance in signage and wayfinding, for, as Alan Fletcher once observed to me apropos of his signage at Stansted Airport outside London, "people look for signs when they are lost, not when they start." For Thomas Manss, who worked with Fletcher at Pentagram when he first came to London "wayfinding has always been close to my heart." It is a discipline, he muses, whose principles do not just apply to signage design but are also very relevant to website design, in empowering the user to find his way around the site. The craft of good wayfinding consists in finding outcomes that are legible, accessible and clear, particularly in cases of hierarchies of signage. "Signage is not a stage for funny jokes or big design ideas," Thomas Manss comments, "as these often only impede the clarity of wayfinding." Context can also influence the way signage operates. This was something Thomas Manss had to reconcile when he was invited in 2004 to design the external and internal exhibition and associated signage for Tate Modern and Tate Britain.

The well-deserved success of Tate Modern was, I believe, in part because it was long planned. In 1985, in presenting an exhibition of drawings by Christopher Wren at the Whitechapel Gallery (of which he was then director) Nicolas Serota spoke of the need to revive the Thames as a London artery, bringing the river back into the heart of the metropolis (a theme also proposed at the time by the architect Richard Rogers). Tate Modern has indeed revived the area south of the Thames across from the City (the house where Wren lived during the building of St Paul's Cathedral is adjacent to Tate Modern). The stunning architectural statement of the former generating station at Bankside, by the architects Herzog and de Meuron, has made the building iconic, just as Stirling's Clore Gallery at the former Tate Gallery (now Tate Britain) revived interest in the museum and its collections some twenty years ago. The opportunity to rehang, redistribute and so revisit the collections with the split across two sites was also a reinvigorating opportunity. The series of temporary exhibitions in both places has raised the status of both. But each site is a major architectural statement, within a highly sensitive visual context.

The Unilever Series: Bruce Nauman
An annual art commission sponsored by Unilever
August Strindberg: Painter Photographer
Damien Hirst's Tate Boat
The fastest, most stylish boat on the river
TATE MEMBERS

"Signage in such a context is in what might be called a servant situation. It has an important job to do," Thomas comments, "but it should not try and create a second personality in competition or conflict with the main one, which in both cases is very strong." What is required in such a case is firstly the choice of a typeface which is legible, efficient and relevant, but not bland, and secondly a context for the typeface which clearly separates it from other visual data. Understanding the limitations posed by the physical setting is an important part of the design concept.

Type was used internally in dark brown or white, either backlit or on glass, with at times a second backing colour which changed as the exhibition programme evolved. Externally, for banners and signs, type was used in dark tones against a pastel background, or in white with a second colour as background.

At first sight it might appear that the constraints of such a complex site only serve to limit the designer's potential. In truth, such constraints, if clearly understood, serve to focus and concentrate the designer's vision towards finding an apt and balanced solution, as was signally the case here.

Beyond
Painting:
Burri
Fontana
Manzoni
MODERN
TATE
NAUMAN
STRINDBERG

MODERN
TATE

The Unilever Series: Bruce Nauman
12 October 2004 – 2 May 2005
The Unilever Series: an annual art commission sponsored by Unilever
Joseph Beuys: Actions,
4 February – 2 May 2005
Supported by Tate International Council

hes, Environments
August Strindberg: Painter, Photographer, Writer
17 February – 15 May 2005
Supported by TATE MEMBERS
MODERN
TATE

The Unilever Series:
Bruce Nauman

The Unilever Series:
Bruce Nauman

articulation

It is the desperate moment when we discover that this empire, which had seemed to us the sum of all wonders, is a formless ruin, that corruption's gangrene has spread too far to be healed by our sceptre, that the triumph over enemy sovereigns has made us the heirs of their long undoing. Only in Marco Polo's accounts was Kublai Khan able to discern, through the walls and towers destined to crumble, the tracery of a pattern so subtle it could escape the termites' gnawing.

Italo Calvino: *Invisible Cities*, translated by William Weaver, 1974

Ostensibly this book describes the cities that Marco Polo visits on his travels and tells Kublai Khan about on his return to the emperor's palace. Ostensibly, since none of the cities seem to have existed at the time – some, constructed from steel, aluminium and plate glass, could never have – and further because the descriptions are not the kind of military or mercantile, political or parochial information that might have been some use to a ruler, but rather evocations of the genius loci of each city. Nor are the accounts arranged following a map (indeed there are no geographical relationships given between the cities at all). The structure is created by the author, who defines eleven categories of city ('cities and desire, cities and memory, cities and the dead, cities and the sky, hidden cities', and so on). A new category is introduced as the final entry in each chapter, giving the book a deliberate, ludic cadence.

Some critics have seen in this quasi-pataphysical exercise nothing but a romantic and literary indulgence – form for form's sake. There is a certain elegiac elegance to the texts, especially in the conversations between Marco Polo and the Khan, but the deeper message of the book is about the failure of quantification to describe quality: it is a subtle attack on certain conventions of Modernism that put the formal and physical values of architecture and urbanism above the human experience: whatever Marco Polo might have known about the fortifications, marketplaces, wealth or political system of the cities would have been of material interest to the Khan, but what the Khan would not have learned was what the cities were like as places to live in. Calvino's systematic presentation only accents the intangible qualities of his inventions.

Designers use systems as well: the grids created in laying out book and magazine pages are a prime example. The carefully considered rules for placing corporate logos on paperwork or in advertising are another, as are the planning and ordering of the structure of levels of information on a website, or the axes of access on a trade fair stand. Such systems exist not just to satisfy the designer's sense of order, but to help the user: the reader should absorb almost unconsciously the difference between headline, sub-head, main text and caption on a magazine page, for example. More importantly, the system chosen should reflect the invisible qualities of the material being designed. To set the main love story in a teen fiction magazine in aggressive sans serif type would be at worst a mistake: to set the entries in a medical dictionary in a florid swash face would be foolhardy.

Designers need systems because design users can read and benefit from them: systems structure communication. But designers need also to understand and convey the emotions, values and purposes of the subject. Designers need to build invisible cities too.

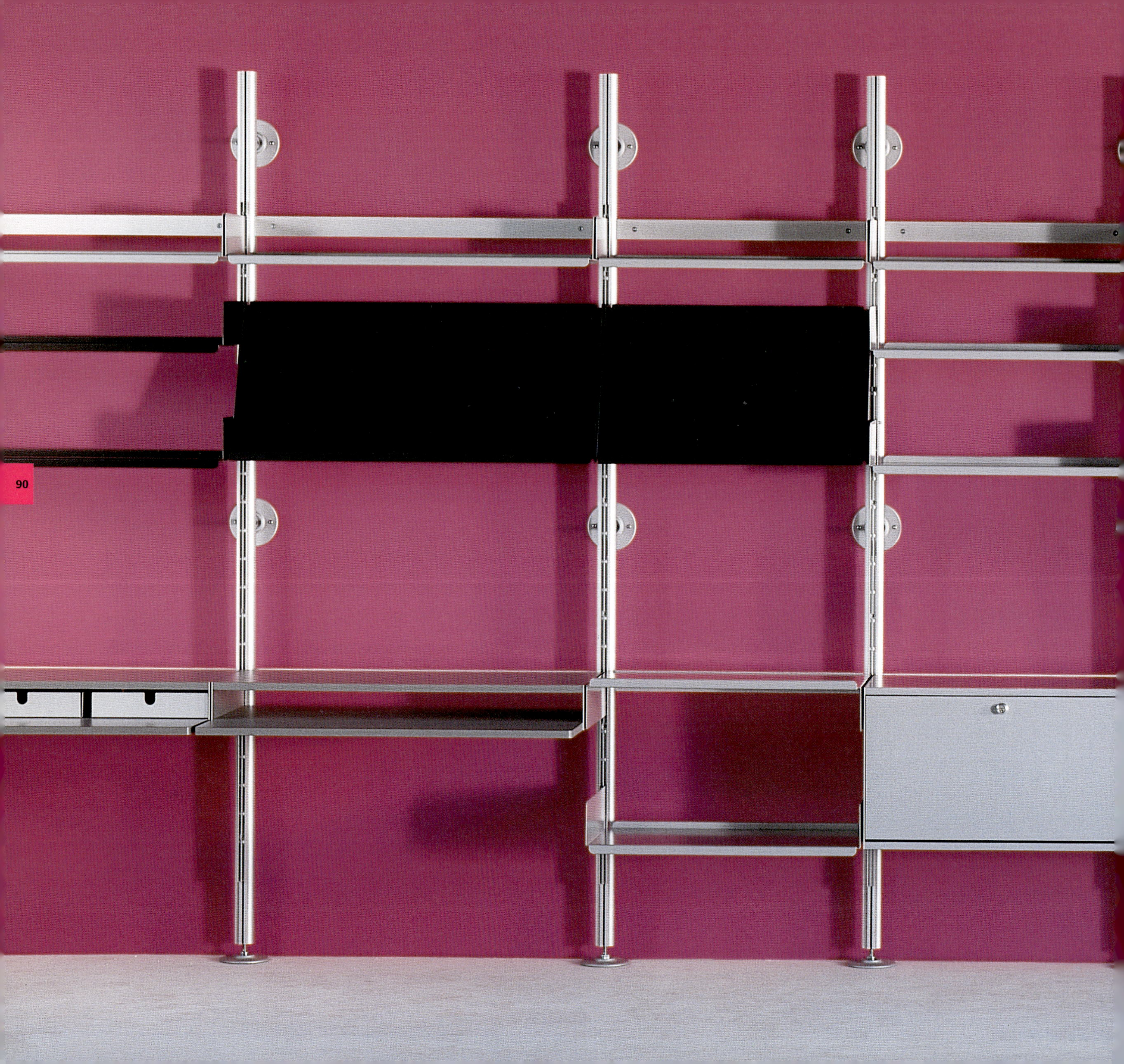

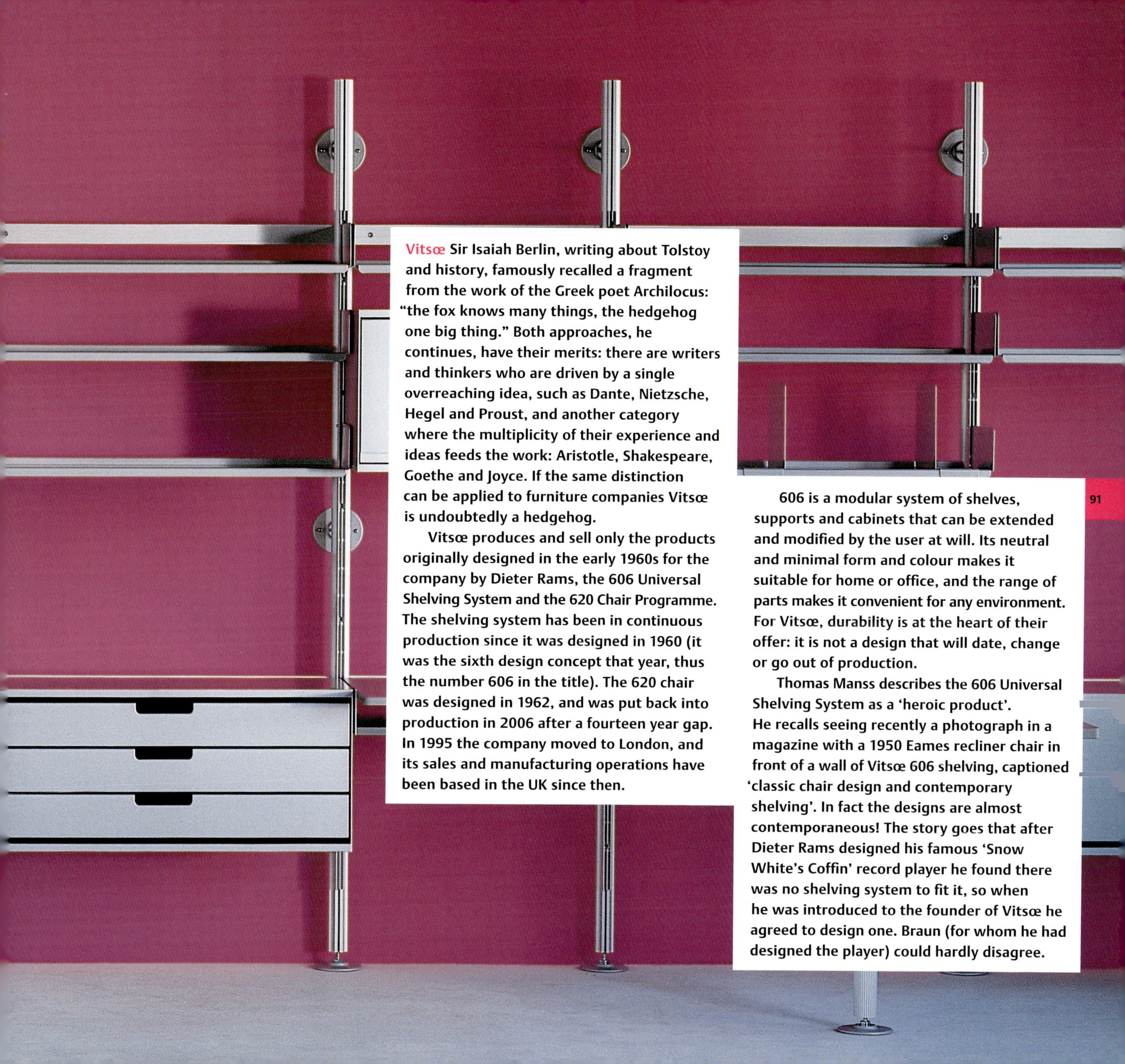

Vitsœ Sir Isaiah Berlin, writing about Tolstoy and history, famously recalled a fragment from the work of the Greek poet Archilocus: "the fox knows many things, the hedgehog one big thing." Both approaches, he continues, have their merits: there are writers and thinkers who are driven by a single overreaching idea, such as Dante, Nietzsche, Hegel and Proust, and another category where the multiplicity of their experience and ideas feeds the work: Aristotle, Shakespeare, Goethe and Joyce. If the same distinction can be applied to furniture companies Vitsœ is undoubtedly a hedgehog.

Vitsœ produces and sell only the products originally designed in the early 1960s for the company by Dieter Rams, the 606 Universal Shelving System and the 620 Chair Programme. The shelving system has been in continuous production since it was designed in 1960 (it was the sixth design concept that year, thus the number 606 in the title). The 620 chair was designed in 1962, and was put back into production in 2006 after a fourteen year gap. In 1995 the company moved to London, and its sales and manufacturing operations have been based in the UK since then.

606 is a modular system of shelves, supports and cabinets that can be extended and modified by the user at will. Its neutral and minimal form and colour makes it suitable for home or office, and the range of parts makes it convenient for any environment. For Vitsœ, durability is at the heart of their offer: it is not a design that will date, change or go out of production.

Thomas Manss describes the 606 Universal Shelving System as a 'heroic product'. He recalls seeing recently a photograph in a magazine with a 1950 Eames recliner chair in front of a wall of Vitsœ 606 shelving, captioned 'classic chair design and contemporary shelving'. In fact the designs are almost contemporaneous! The story goes that after Dieter Rams designed his famous 'Snow White's Coffin' record player he found there was no shelving system to fit it, so when he was introduced to the founder of Vitsœ he agreed to design one. Braun (for whom he had designed the player) could hardly disagree.

When Thomas Manss started working with Vitsœ in 1994 the company's showroom doubled as the owner's London base (so by appointment only): they now have a dedicated showroom in London's Wigmore Street, a growing customer sector in the USA and internationally, and have recently gone into the Scandinavian market. The first request was to look at the logo and identity. Since their arrival in London, Vitsœ had asked Addison, Neville Brody and others to update it. Wolfgang Schmidt had designed the original identity in the 1960s: Thomas Manss felt that the right approach was to go back to the original identity; after all continuity was part of the company's principles. The agency has since designed literature for Vitsœ, their website and stands at the 100% Design fair in London.

"The 606 shelving system is a premium product," Thomas explains, "and so we needed to convince new users that the premium was worth it over time, because of the flexibility and robustness of the product. If you had bought a system thirty years ago and now needed a dozen more shelves, they would still fit the system and match the originals." One way of conveying this was to use the existing customer base to promote the product: they were often enthusiasts for Vitsœ and 606, and so, for example, for the brochure Manss used photographs of Vitsœ aficionados such as Ross Lovegrove, Dada Rogers, Jeremy King and the architects at Future Systems, Jan Kaplicky and Amanda Levete, among others, together with their narratives of using the system. This put human faces – and successful role models – onto an otherwise formal and abstract product.

For the website Mark Adams of Vitsœ had originally gone to a team of web experts but was none too happy with the results, all flying objects and visual tricks, so he asked Thomas to help. "We agreed that Vitsœ wasn't a flying object sort of company, but what it needed was a website that would educate customers about the values of the product. We designed one of the early content managed sites, to which Vitsœ could add images of new installations as they wanted." In fact, the images do most of the work, and Vitsœ find the website offers them a silent dialogue with potential customers: they can track where people enter and leave the site, what they stop to look at, and can continuously adjust the content accordingly.

For the 100% Design show the brief was rather different: how to present a forty-year old product among a host of ultra-new ideas and concepts. Thomas's solution was shocking: shocking pink, that is, using a boldly contemporary colour to highlight the versatility, endurance and relevance of what Vitsœ has been systematically offering and is still on offer.

VITSŒ
606 Universal Shelving System™

The art of living better

Dieter Rams, 1971

It's a funny old world. The realisation is dawning that just spending more time earning more money to buy more stuff that gives transient gratification is not necessarily the route to eternal happiness. The developed world is moving from scarcity to surfeit. The backlash must come.

Can we honestly continue to use everything for such a short period of time and then not feel a pang of guilt when we have to throw it away? That plastic cup for a single drink of water on the aeroplane; that 18 month-old printer that is cheaper to replace than repair; all of that left-over packaging from your trip to the supermarket; and that obsolete three year-old computer. What happens if/when the entire world adopts these habits? According to current thinking, we will need three planets to sustain us.

In 1987 the Brundtland Commission defined sustainability:

"Sustainable development is development that meets the needs of the present without compromising the ability of future generations to meet their own needs."

Yet everything we do in our lives seems to have a negative impact on the world around us. So, what are we to do?

How about creating products that are avowedly long-term in their outlook? Products that do not strive for built-in obsolescence but prefer to be discreet, adaptable and faithful servants in the face of a turbulent world. Products that minimise their inevitable impact on the world's environment and resources by being useful for as long as possible.

This was Vitsœ's proposition in 1959: to eschew fashion whilst creating products that would be the neutral canvas on which to paint your colourful life. After almost half a century our resolve is stronger than ever: more of us must learn the art of living better, with less, that lasts longer.

'loristic Affinities', /itsœ, London.)esigned and :urated by he landscape :rchitects (and 06 owners), &L Gibbons, *Betula pendula* silver birch) was :ombined with 06 to show two ioneering species n harmony

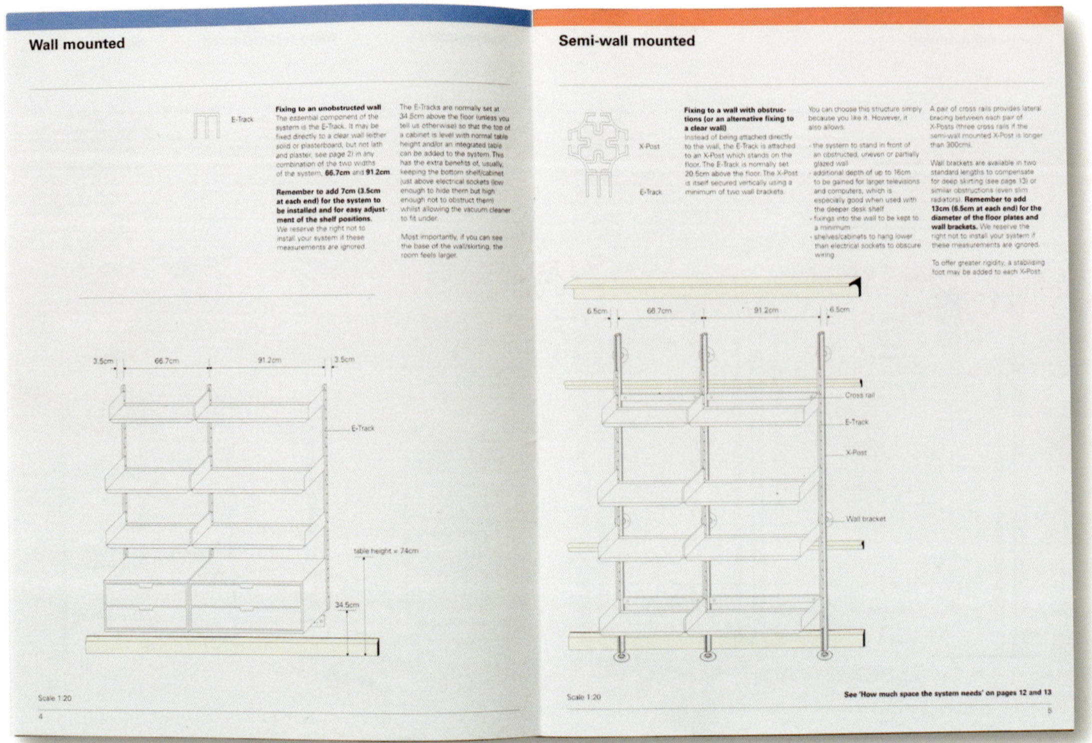

Wall mounted

E-Track

Fixing to an unobstructed wall
The essential component of the system is the E-Track. It may be fixed directly to a clear wall (either solid or plasterboard, but not lath and plaster, see page 2) in any combination of the two widths of the system, **66.7cm** and **91.2cm**.

Remember to add 7cm (3.5cm at each end) for the system to be installed and for easy adjustment of the shelf positions. We reserve the right not to install your system if these measurements are ignored.

The E-Tracks are normally set at 34.5cm above the floor (unless you tell us otherwise) so that the top of a cabinet is level with normal table height and/or an integrated table can be added to the system. This has the extra benefits of, usually, keeping the bottom shelf/cabinet just above electrical sockets (low enough to hide them but high enough not to obstruct them) whilst allowing the vacuum cleaner to fit under.

Most importantly, if you can see the base of the wall/skirting, the room feels larger.

Scale 1:20

4

Semi-wall mounted

X-Post
E-Track

Fixing to a wall with obstructions (or an alternative fixing to a clear wall)
Instead of being attached directly to the wall, the E-Track is attached to an X-Post which stands on the floor. The E-Track is normally set 20.5cm above the floor. The X-Post is itself secured vertically using a minimum of two wall brackets.

You can choose this structure simply because you like it. However, it also allows:

- the system to stand in front of an obstructed, uneven or partially glazed wall
- additional depth of up to 16cm to be gained for larger televisions and computers, which is especially good when used with the deeper desk shelf
- fixings into the wall to be kept to a minimum
- shelves/cabinets to hang lower than electrical sockets to obscure wiring

A pair of cross rails provides lateral bracing between each pair of X-Posts (three cross rails if the semi-wall mounted X-Post is longer than 300cm).

Wall brackets are available in two standard lengths to compensate for deep skirting (see page 13) or similar obstructions (even slim radiators). **Remember to add 13cm (6.5cm at each end) for the diameter of the floor plates and wall brackets.** We reserve the right not to install your system if these measurements are ignored.

To offer greater rigidity, a stabilising foot may be added to each X-Post.

Scale 1:20

See 'How much space the system needs' on pages 12 and 13

5

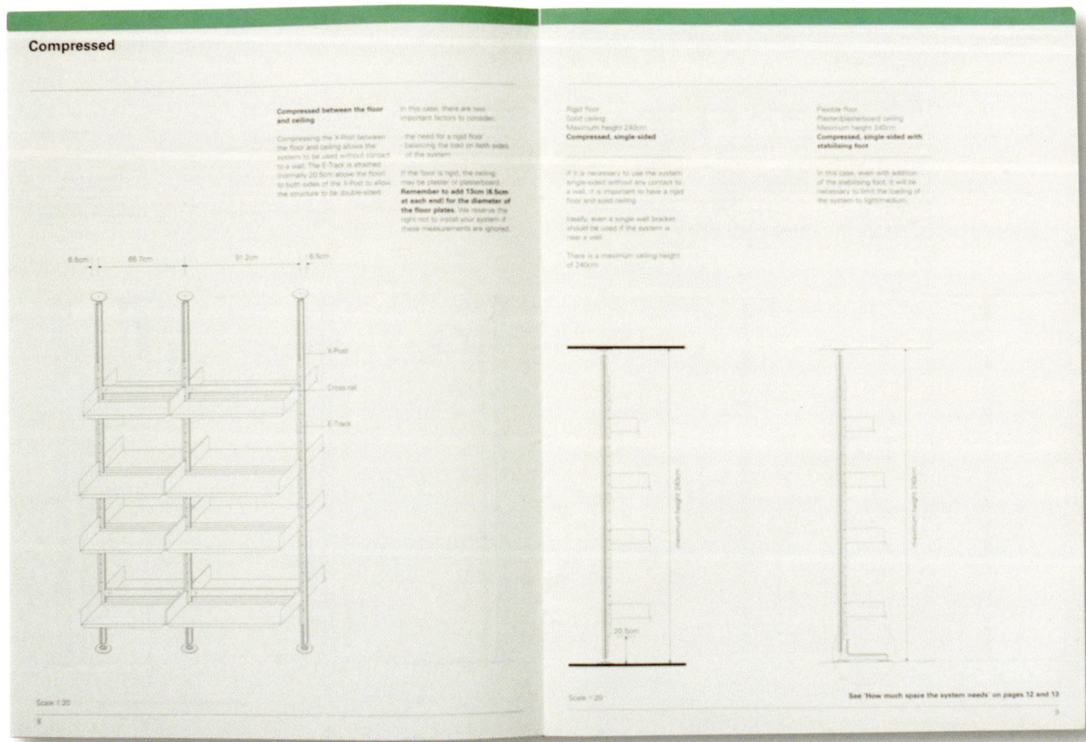

Compressed

Compressed between the floor and ceiling

Remember to add 13cm (6.5cm at each end) for the diameter of the floor plates.

Compressed, single sided

Compressed, single-sided with stabilising foot

See 'How much space the system needs' on pages 12 and 13

VITSŒ
606 Universal Shelving System
Installation and Beyond

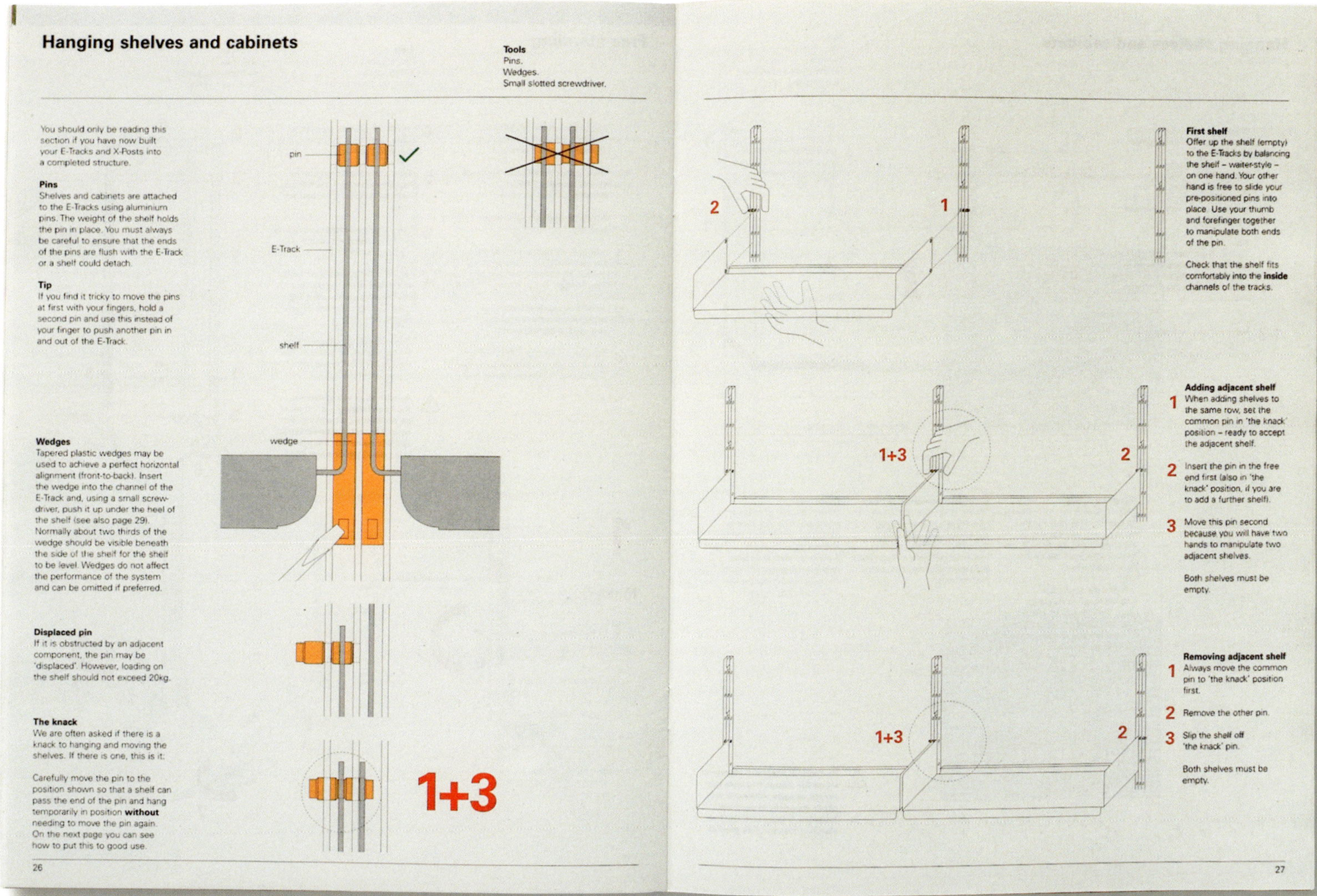

Hanging shelves and cabinets

Tools
Pins.
Wedges.
Small slotted screwdriver.

You should only be reading this section if you have now built your E-Tracks and X-Posts into a completed structure.

Pins
Shelves and cabinets are attached to the E-Tracks using aluminium pins. The weight of the shelf holds the pin in place. You must always be careful to ensure that the ends of the pins are flush with the E-Track or a shelf could detach.

Tip
If you find it tricky to move the pins at first with your fingers, hold a second pin and use this instead of your finger to push another pin in and out of the E-Track.

Wedges
Tapered plastic wedges may be used to achieve a perfect horizontal alignment (front-to-back). Insert the wedge into the channel of the E-Track and, using a small screwdriver, push it up under the heel of the shelf (see also page 29). Normally about two thirds of the wedge should be visible beneath the side of the shelf for the shelf to be level. Wedges do not affect the performance of the system and can be omitted if preferred.

Displaced pin
If it is obstructed by an adjacent component, the pin may be 'displaced'. However, loading on the shelf should not exceed 20kg.

The knack
We are often asked if there is a knack to hanging and moving the shelves. If there is one, this is it:

Carefully move the pin to the position shown so that a shelf can pass the end of the pin and hang temporarily in position **without** needing to move the pin again. On the next page you can see how to put this to good use.

26

First shelf
Offer up the shelf (empty) to the E-Tracks by balancing the shelf – waiter-style – on one hand. Your other hand is free to slide your pre-positioned pins into place. Use your thumb and forefinger together to manipulate both ends of the pin.

Check that the shelf fits comfortably into the **inside** channels of the tracks.

Adding adjacent shelf
1 When adding shelves to the same row, set the common pin in 'the knack' position – ready to accept the adjacent shelf.
2 Insert the pin in the free end first (also in 'the knack' position, if you are to add a further shelf).
3 Move this pin second because you will have two hands to manipulate two adjacent shelves.

Both shelves must be empty.

Removing adjacent shelf
1 Always move the common pin to 'the knack' position first.
2 Remove the other pin.
3 Slip the shelf off 'the knack' pin.

Both shelves must be empty.

27

wallpaper*
to die for
regal rio / palm springs eternal
sacred spaces / metal urges
cool in kinshasa / fairy tale fashion
juicy jewellery / immortal morsels

SPECIAL TRAVEL ISSUE – PART 1 OF OUR 2003 GUIDE
wallpaper
*
odyssey or oddity?
wallpaper* sizes up some urban myths
+ our 108-page german survey

VITSŒ
For our brochure, planning guide and price list, please contact us:
Telephone +44 (0)20 7354 8444
Fax +44 (0)20 7354 9888
Email orange@vitsoe.com
Internet www.vitsoe.com

606 Universal Shelving System™
designed by Dieter Rams in 1960
and produced continually ever since
Timeless, movable and constantly evolving

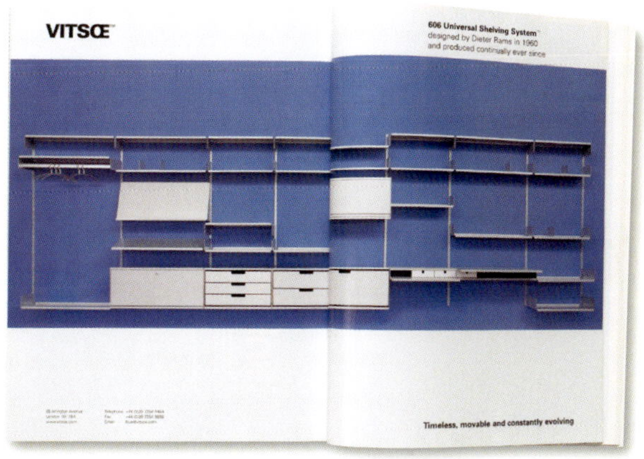
VITSŒ
606 Universal Shelving System™
designed by Dieter Rams in 1960
and produced continually ever since
Timeless, movable and constantly evolving

VITSOE
EXIT
100
DIFFA
AIDS

74
72
VITSŒ
Now

crossing borders

"The species of the Fairing are more divided by their sense of time than anything else. We Dwellers, being who and what we are, naturally encompass as much of the spectrum of chronosense as we are able, covering most of it. I exclude the machine-Quick." A hesitation "You still abhor those, I take it?" "Yes, we most certainly do!" the colonel exclaimed. "Positively persecuted," Fassin said. The speaker was an ancient Sage called Jundriance.

Iain M. Banks: *The Algebraist*, 2004

Iain M. Banks' science-fiction novel *The Algebraist* sits apart from some of his other novels about the Culture, to consider how a species, humanoid but living for centuries, the Quick, interact with a species, the Dwellers, or the Slow, whose lifetimes are counted in millennia. The Dwellers are quaint, at times comic and yet complex and powerful, with a world vision from within their gas giant planets that operates on a timescale quite different from the Quick, who are caught up in the rivalries of different military and political factions, despite living among a host of different species. The book shows Banks' endless inventive ingenuity as well as his extensive skills with language, expression and nuance. The narrative, a search for a document that may well never have existed, against a background of interplanetary warfare, is framed by this crossover of cultures, which gives depth and relevance to what might otherwise be an adventure story *tout court*.

Science fiction is a beguiling medium, in that from the author's point of view everything, from bug-eyed monsters (upwards and downwards) can be invented or invoked to assist the narrative. But Banks generates a sense of logic in his work without such evident devices, through the coherence of the writing itself. Banks' writing, whether sci-fi or conventional, benefits from – or rather takes maximum advantage of – a certain kind of postModern position, in which boundaries exist so as to be transgressed.

It is perhaps too soon to measure the real contribution of the postModern vision, for various reasons. Firstly there are the constraints and exigencies of the Cold War, which for political or ideological reasons preserved the shelf-life of Modernism beyond its sell-by date. Secondly there are the immediate excesses of early postModernism, particularly in architecture, which led to a flurry of indulgent decoration or rather decorativisme. Thirdly, and perhaps in the long term more positively, there was independently a technological revolution in the pre-press of visual material, followed in short order by the arrival of the World Wide Web, which destabilised the concepts of order and hierarchies of content that had prevailed to date.

Design too has benefited from this sea-change, allowing designers to explore new areas of activity – advertising for example – and above all not to feel that they are restricted in their choices through a professional castesystem ("I am a graphic designer, you are an interior designer, she is a product designer," etc). Freedom, of course, brings responsibility, but it is still freedom.

National Portrait Gallery The contemporary exhibition catalogue has moved across from being an illustrated list of the work on show together with a few essays on the topic in two directions: one towards being more like an independent book, containing enough information to stand alone, and the other, on the contrary, towards being even more firmly integrated into the exhibition, being part of the marketing strategy of the show and one of the principal sources of marketing revenue. It was such a slightly schizophrenic publication that Thomas Manss was asked to design for the National Portrait Gallery.

The NPG, as it is known, is London's storehouse of historical portraits of the great, good and notorious men and women of Britain, located next to the National Gallery in Trafalgar Square. In recent years, the gallery has increasingly become interested in portrait photography, and it was an exhibition entitled *Face of Fashion* for which Thomas Manss was asked to design the catalogue. The exhibition showed the work of five photographers or photographic teams, Corinne Day, Steven Klein, Mario Sorrenti, Mert Alas and Marcus Piggott, and Paolo Roversi, each famous for bringing unconventional styling to their editorial or advertising portraits, often of celebrities: Corinne Day's images of Kate Moss, for example, being very well known. The exhibition opened in February 2007 and ran until May.

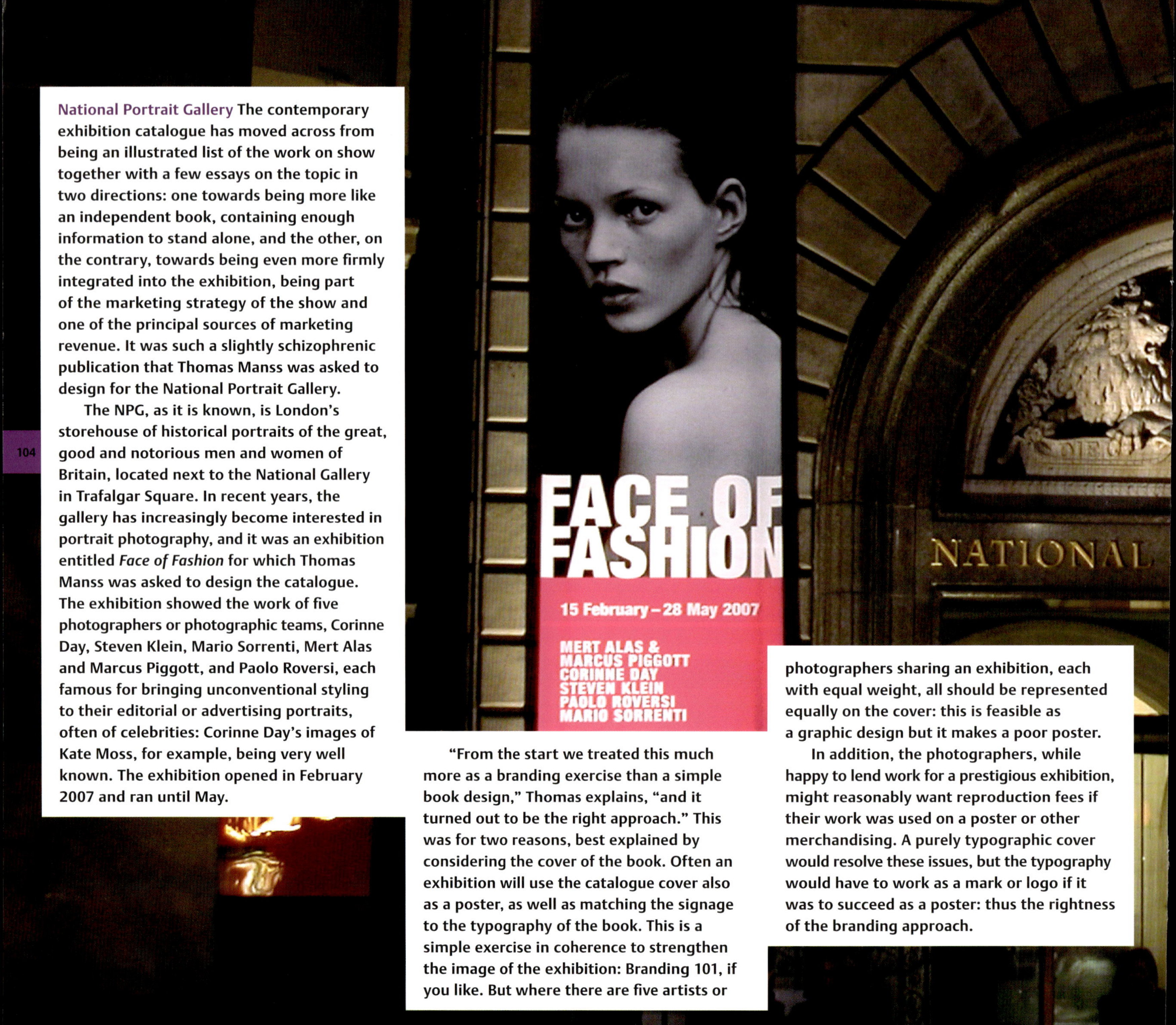

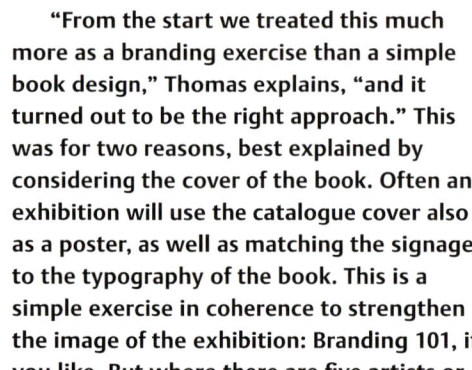

"From the start we treated this much more as a branding exercise than a simple book design," Thomas explains, "and it turned out to be the right approach." This was for two reasons, best explained by considering the cover of the book. Often an exhibition will use the catalogue cover also as a poster, as well as matching the signage to the typography of the book. This is a simple exercise in coherence to strengthen the image of the exhibition: Branding 101, if you like. But where there are five artists or photographers sharing an exhibition, each with equal weight, all should be represented equally on the cover: this is feasible as a graphic design but it makes a poor poster.

In addition, the photographers, while happy to lend work for a prestigious exhibition, might reasonably want reproduction fees if their work was used on a poster or other merchandising. A purely typographic cover would resolve these issues, but the typography would have to work as a mark or logo if it was to succeed as a poster: thus the rightness of the branding approach.

Portrait
Gallery
FACE OF
FASHION
15 February – 28 May 2007
MERT ALAS &
MARCUS PIGGOTT
CORINNE DAY
STEVEN KLEIN
PAOLO ROVERSI
MARIO SORRENTI
GAP
RTRAIT GALLERY

FACE OF FASHION

Making the title into a logo or brand also opened up other possibilities for merchandising: it could be used on carrier-bags, T-shirts, mugs, notebooks and the other paraphernalia which museums and galleries now offer to visitors as mementoes of their exhibitions. Here again the reproduction rights issue would not arise. The typography chosen for the book cover was Neue Helvetica Black Condensed, a strong and compact sans serif, printed vertically in pink as two lines on a dark background.

In discussing the project with staff at the NPG, Thomas Manss discovered that one of the logics behind the exhibition was to attract younger visitors to the NPG: their current visitor profile was mainly of people over forty-five years of age. Why not, therefore, Thomas Manss suggested, use the *Face of Fashion* brand that had been created to front a marketing exercise aimed at younger visitors. There was not a large marketing budget, but the NPG agreed. The result was both a range of merchandise that was appropriate for a younger market, and more importantly, a dedicated website, so that anyone googling 'fashion' or 'Kate Moss' would find it without going through the main NPG portal, though it was linked to that as well.

The NPG also agreed to fund a poster, using one of Mario Sorrenti's monochrome portraits of Kate Moss, seen as being the most iconic image in the show for the target market. On the website the visitor could select a pixel, which would entitle them, when they visited the exhibition, to a free copy of the poster itself: a simple and original incentivisation mechanism. Visitors to the site could also download the poster as wallpaper for a mobile phone screen, or a selection of photographs from the exhibition as a computer screensaver.

"The way all these aspects were brought together was only possible because of the branding approach we adopted from the start," Thomas Manss notes. In other words, because the agency was willing to cross boundaries, to apply a discipline in an unexpected context, the result was an extension of the possibilities offered initially, into a new strategy for reaching a desired market. Not rocket science, not even algebra, just simple lateral thinking.

Face of Fashion Catalogue
Double page from chapter:
Corinne Day

MARIO SORRENTI

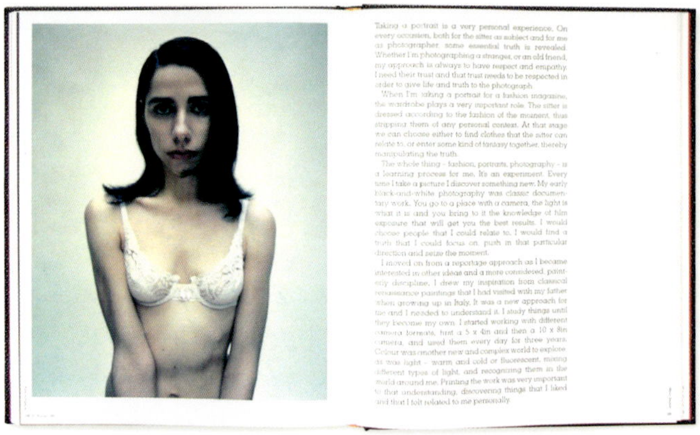
Taking a portrait is a very personal experience. On every occasion, both for the sitter as subject and for me as photographer, some essential truth is revealed. Whether I'm photographing a stranger, or an old friend, my approach is always to have respect and empathy. I need their trust and that trust needs to be respected in order to give life and truth to the photograph.
When I'm taking a portrait for a fashion magazine, the wardrobe plays a very important role. The sitter is dressed according to the fashion of the moment, thus stripping them of any personal context. At that stage we can choose either to find clothes that the sitter can relate to, or enter some kind of fantasy together, thereby manipulating the truth.
The whole thing – fashion, portraits, photography – is a learning process for me. It's an experiment. Every time I take a picture I discover something new. My early black-and-white photography was classic documentary work. You go to a place with a camera, the light is what it is and you bring to it the knowledge of film exposure that will get you the best results. I would choose people that I could relate to. I would find a truth that I could focus on, push in that particular direction and seize the moment.
I moved on from a reportage approach as I became interested in other ideas and a more considered, painterly discipline. I drew my inspiration from classical renaissance paintings that I had visited with my father when growing up in Italy. It was a new approach for me and I needed to understand it. I study things until they become my own. I started working with different camera formats, first a 5 x 4in and then a 10 x 8in camera, and used them every day for three years. Colour was another new and complex world to explore, as was light – warm and cold or fluorescent, mixing different types of light, and recognizing them in the world around me. Printing the work was very important to that understanding, discovering things that I liked and that I felt related to me personally.

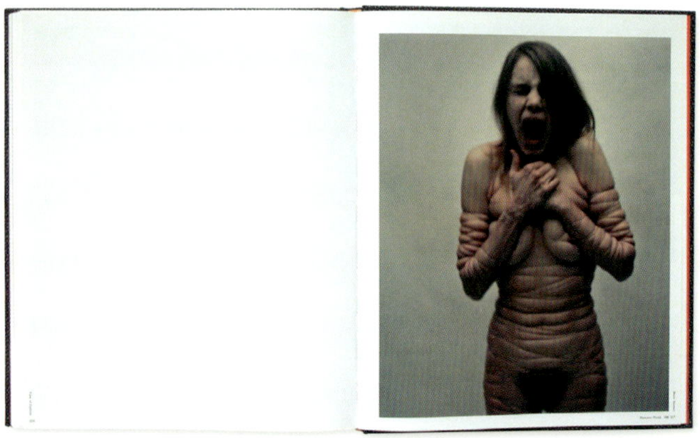

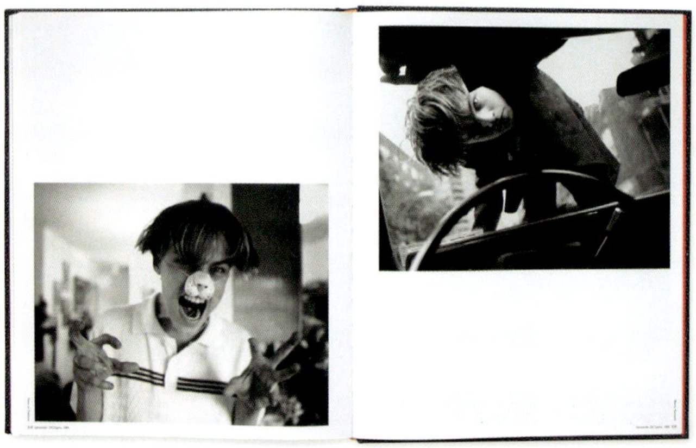

110

Face of Fashion Catalogue
Chapter: Mario Sorrenti

National Portrait Gallery – Face of Fashion
National Portrait Gallery
15 February – 28 May 2007
FACE OF FASHION
MERT ALAS & MARCUS PIGGOTT
CORINNE DAY
STEVEN KLEIN
PAOLO ROVERSI
MARIO SORRENTI
ENTER

National Portrait Gallery – Face of Fashion
FACE OF FASHION
MERT ALAS & MARCUS PIGGOTT
CORINNE DAY
STEVEN KLEIN
PAOLO ROVERSI
MARIO SORRENTI
YOUR VISIT
NEW YORK COMPETITION
DESIGN YOUR OWN
DOWNLOADS
PRESS
CATALOGUE
National Portrait Gallery
15 February – 28 May 2007
Mert Alas and Marcus Piggott
Mert Alas (b.1971) and Marcus Piggott (b.1970) formed their creative partnership in the late 1990s. Based in London and Ibiza, they are best known for their advertising work for luxury brands and their highly polished, but somewhat uneasy, editorial work for magazines such as Pop, W and Vogue. Their characteristic style is a skilful mix of the old-school glamour of 1930s Hollywood portraiture with a futuristic, synthetic, post-production gloss. What might otherwise be read as simple flattery or a concession to beauty is pushed to appear almost artificial. Often humorous, their work explores stereotypes of female identity through the use of role play and fantasy.
Click here to download an audio tour and find out more about Mert Alas & Marcus Piggott and the works featured in the exhibition.
After Dark, Berlin, 2004
© Mert Alas and Marcus Piggott

National Portrait Gallery – Face of Fashion
FACE OF FASHION
MERT ALAS & MARCUS PIGGOTT
CORINNE DAY
STEVEN KLEIN
PAOLO ROVERSI
MARIO SORRENTI
YOUR VISIT
NEW YORK COMPETITION
DESIGN YOUR OWN
DOWNLOADS
PRESS
CATALOGUE
National Portrait Gallery
15 February – 28 May 2007
Corinne Day
Corinne Day (b.1965) is a British photographer whose influence on the style and perception of photography in the early 1990s has been immense. A former model and self-taught photographer, she brought a more hard-edged documentary look to fashion image-making. She is known for forming close relationships with many of her sitters (most famously Kate Moss), which have resulted in candid and intimate portraits. Her approach, as illustrated within the fashion magazines of the 1990s, came to be known as 'grunge' and grew into an international style.
Day continues to take photographs for fashion magazines and has worked on international advertising campaigns for designers such as Vivienne Westwood and Gap.
Click here to download an audio tour and find out more about Corinne Day and the works featured in the exhibition.
Kirsten, 2000
© Corinne Day, Courtesy of the Artist and Gimpel Fils, London

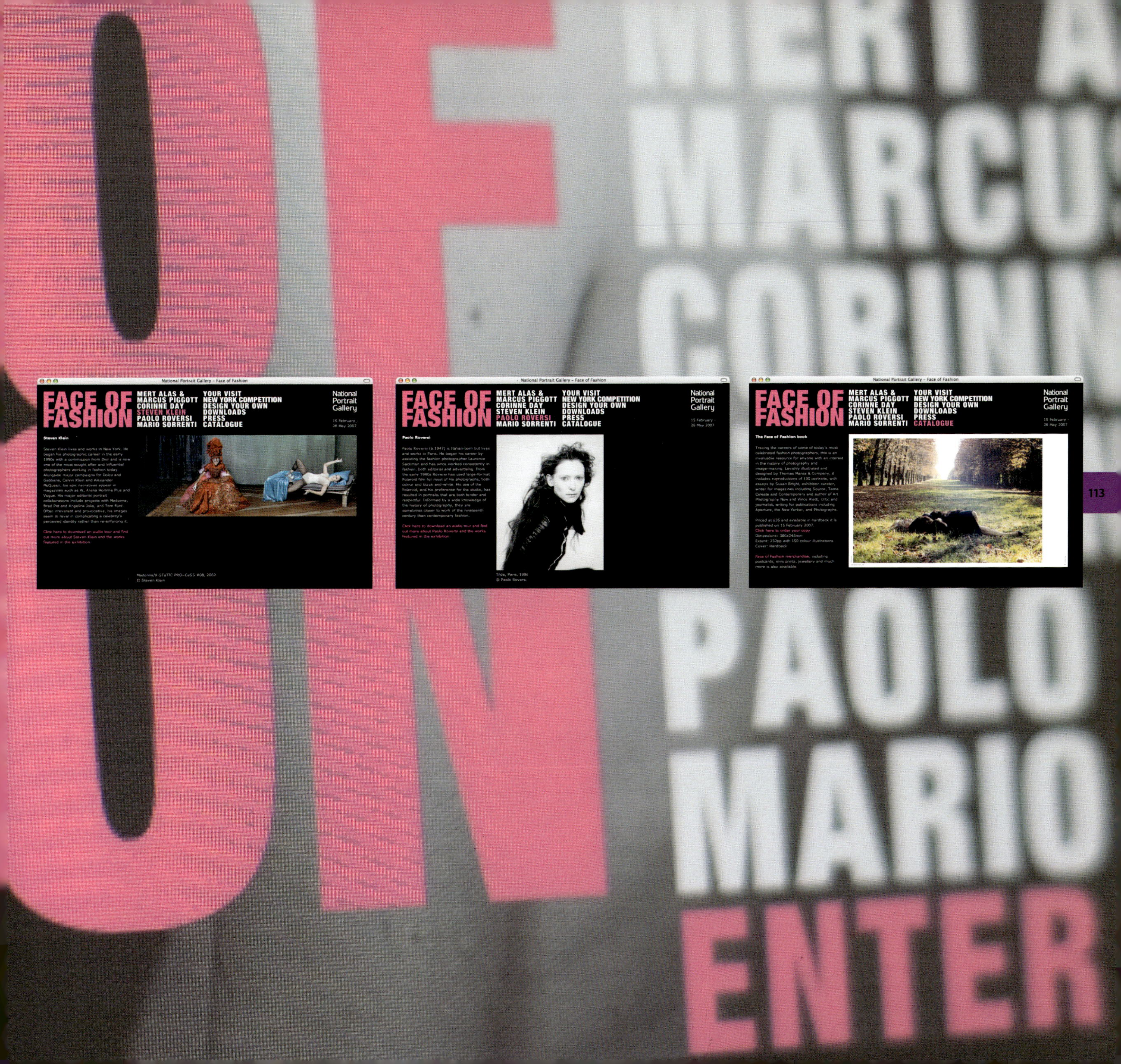

FACE OF FASHION
MERT ALAS &
MARCUS PIGGOTT
CORINNE DAY
STEVEN KLEIN
PAOLO ROVERSI
MARIO SORRENTI
YOUR VISIT
NEW YORK COMPETITION
DESIGN YOUR OWN
DOWNLOADS
PRESS
CATALOGUE
National Portrait Gallery
PAOLO
MARIO
ENTER

FACE OF FASHION
MERT ALAS &
MARCUS PIGGOTT
CORINNE DAY
STEVEN KLEIN
PAOLO ROVERSI
MARIO SORRENTI

FACE OF FASHION

telling stories

The guard blew his whistle and waved his flag – how weighted with ritual have the railways in their brief century become! – and the train crawled from the little station. The guard walked alongside through the snowflakes, wistful for that jump-and-swing at an accelerating van that is the very core of the mystery of guarding trains. But the train continued to crawl. Sundry footballers in a glass box, some with legs swung high in the air, stood immobile to watch its departure.

Michael Innes: *Appleby's End*, 1945

John Appleby, Michael Innes' detective hero, is involved, in *Appleby's End* in a murder mystery, a romance and a farce, which ends with a reference to the ending of *Tristram Shandy*. This is appropriate given the endless and entertaining literary divagations that are scattered through the text. Inspector Appleby is sent down from London to investigate odd goings on in a place called Snarl, but ends up at Dream Manor, a house full of eccentrics, both alive and dead, uncovering a history – supposedly – of lost wills and missing heirs, and falling in love as well. "There are two things we must be doing," he explains at one point, "getting to the heart of the mystery – and getting away again."

Detective stories, at least in the classic English genre, are structured and anticipatory narratives. Unlike the thriller, in which the author can summon up a new situation or challenge at will, the detective story needs to contain all the information for the resolution of the mystery within it, so that the astute reader can, with luck, uncover the murderer ahead of the final dénouement. The author's skill lies in misleading the reader, but doing so honestly, so that it is only at the end that the reader is aware of the significance of what the dog did in the night-time.

Narrative and anticipation are often key elements in a design project. In exhibition and trade fair work there is a physical journey, through the museum or the stand, that can articulate this sense of learning and discovery. Advertising, especially time-based work such as television commercials, can also make effective use of narrative devices. But there are other, subtler occasions in which telling a story, however indirectly, or taking the user on a journey, even if not physically, embeds a project in the user's understanding.

The mobile phone company Orange was an immediate success because its name, logo and identity (created by Wolff Olins) addressed the consumer's interests, and did not, unlike its competitors at the time, such as Vodafone, merely reflect the fact that it was a phone company. In the same way Wozniak and Jobs' success with the Apple Mac was due to their looking not at the performance of the machine they built but at what the user wanted to have. Both these standpoints invite the user on a journey. So too for Thomas Manss, narrative is an essential part not just of the conception but the execution of a design: "Thomas Manss is graphic designer who is not a graphic designer but a teller of tales and fabricator of fables in the best tradition of the German philosophers who talked too much and never got anything done," as he self-deprecatingly says. But the power of narrative and the skilled way it is put to work is very evident in much of the agency's work.

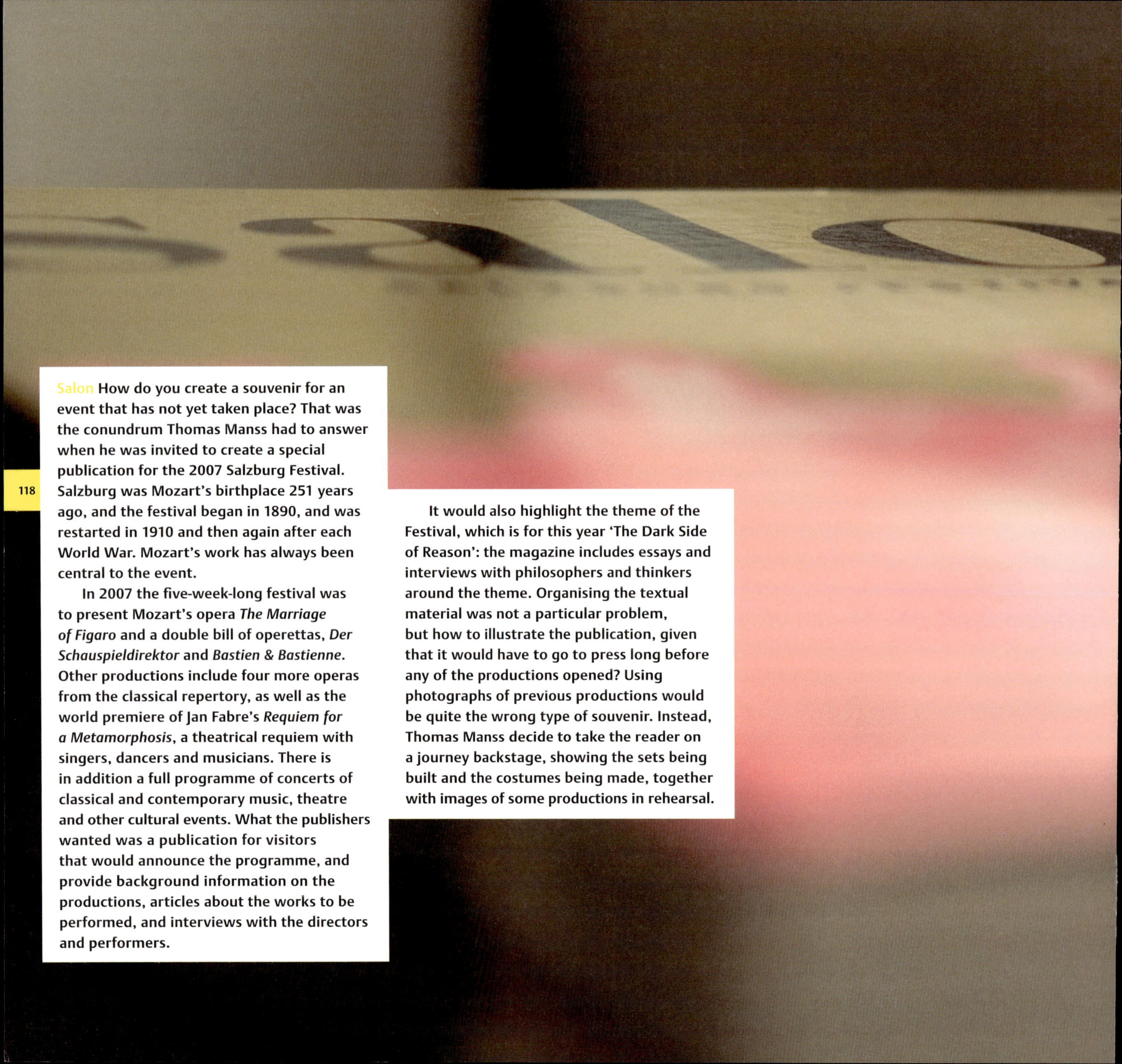

Salon How do you create a souvenir for an event that has not yet taken place? That was the conundrum Thomas Manss had to answer when he was invited to create a special publication for the 2007 Salzburg Festival. Salzburg was Mozart's birthplace 251 years ago, and the festival began in 1890, and was restarted in 1910 and then again after each World War. Mozart's work has always been central to the event.

In 2007 the five-week-long festival was to present Mozart's opera *The Marriage of Figaro* and a double bill of operettas, *Der Schauspieldirektor* and *Bastien & Bastienne*. Other productions include four more operas from the classical repertory, as well as the world premiere of Jan Fabre's *Requiem for a Metamorphosis*, a theatrical requiem with singers, dancers and musicians. There is in addition a full programme of concerts of classical and contemporary music, theatre and other cultural events. What the publishers wanted was a publication for visitors that would announce the programme, and provide background information on the productions, articles about the works to be performed, and interviews with the directors and performers.

It would also highlight the theme of the Festival, which is for this year 'The Dark Side of Reason': the magazine includes essays and interviews with philosophers and thinkers around the theme. Organising the textual material was not a particular problem, but how to illustrate the publication, given that it would have to go to press long before any of the productions opened? Using photographs of previous productions would be quite the wrong type of souvenir. Instead, Thomas Manss decide to take the reader on a journey backstage, showing the sets being built and the costumes being made, together with images of some productions in rehearsal.

For example, the production design of Tchaikovsky's *Eugene Onegin* is based on Russian Social Realist photography, so an article on this theme is included. For Weber's *Der Freischütz*, the costumes are based on the works of the American Pop artist Duane Hanson, who made life-size and life-like figures of ordinary (and often obese) Americans. The costume makers used the same technique of kapok-filled elements to create oversize and overweight characters. This is an interesting story in itself, but it also underlines the theme of the Festival being both historic and contemporary. It also introduces the readers to the people who make the Festival work, both in terms of the direction and the execution of events and performances.

The publication is in a large format, which helps accommodate the two languages used, German and English ("we had a bit of a song and dance choosing the typestyles for that," Thomas remarks.) He also insisted on having a photographer take new portrait photographs of the directors and performers who are featured in the magazine.

The publication can be described as a magazine, though it does not have a magazine's structure of editorials, feature articles and regular columns. But the publication does need to offer the reader the change of pace and perspective that a magazine does, so while concentrating on the main theme of the Festival itself, there are what might be termed bonus pieces – an article on a woman with absolute hearing, who can write down a bird's song in musical notation, or a series of images of the crystalline effects of playing different pieces of music or sounds to a container of water. While following a common typographic system which could best be described as sober but joyous (all type is black to emphasise the images), each main feature seeks to have an individual character and feel. This both gives a change of pace, and also makes each article into an individual journey.

golden
voices
ANNA
NETREBKO
salon

Onegin

Recherchen vor Ort: Auf zwei Reisen durch die Ukraine sammelte Mark Weeger Inspirationen für die Kostüme zu *Eugen Onegin*. Auf dem Fluss Bjeppa mit dem Schiff von Kiew bis zum schwarzen Meer. Die dabei entstandenen Momentaufnahmen zeigen Solisten einer Welt im Wandel.

Research on location: during two trips to Ukraine, Mark Weeger went in search of inspiration for the costumes for *Eugene Onegin*. The snapshots captured on a journey by ship from Kiev to the Black Sea on the river Bjeppa show soloists in a changing world.

Kompositionen des Alltäglichen: Schon Tschaikowski wollte der Realität einen Spiegel vorhalten.
Compositions of ordinary life: Tchaikovsky wanted to hold a mirror up to reality.

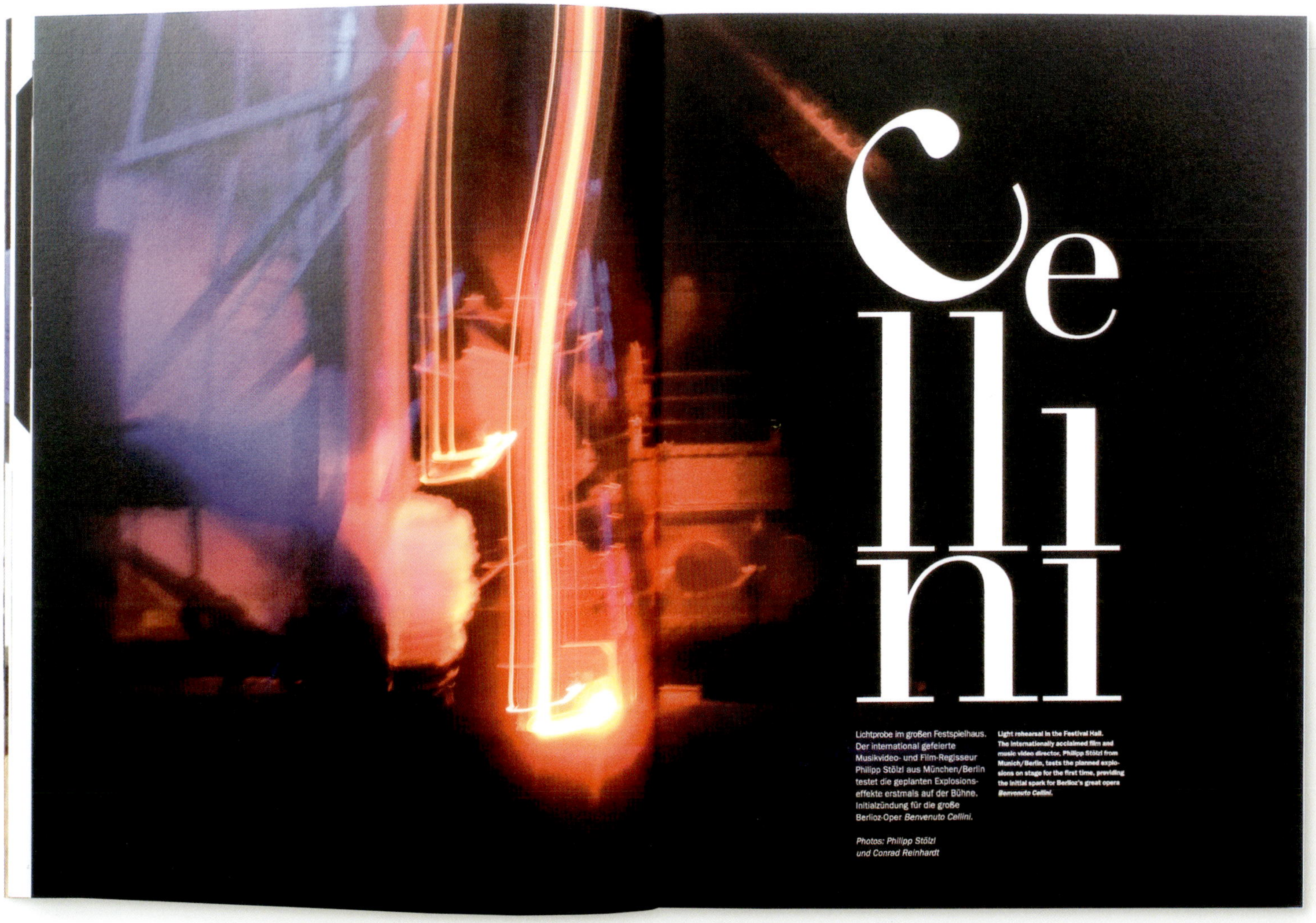

Cellini

Lichtprobe im großen Festspielhaus. Der international gefeierte Musikvideo- und Film-Regisseur Philipp Stölzl aus München/Berlin testet die geplanten Explosionseffekte erstmals auf der Bühne. Initialzündung für die große Berlioz-Oper *Benvenuto Cellini*.

Photos: Philipp Stölzl und Conrad Reinhardt

Light rehearsal in the Festival Hall. The internationally acclaimed film and music video director, Philipp Stölzl from Munich/Berlin, tests the planned explosions on stage for the first time, providing the initial spark for Berlioz's great opera *Benvenuto Cellini*.

frei·schütz

Impression aus der Herrenschneiderei: Ganzkörper-Wattons individuell für die Sänger des *Freischütz*-Chors. Allesamt sind zu schlank für das Kostümkonzept, das sich an den Skulpturen von Duane Hanson orientiert.

Photos: Karin Orell & Mato/Leica

An impression from the men's tailoring workshop: individual full body suits for the singers in the *Freischütz* choir. They were all to thin for the costuming concept, which is oriented towards Duane Hanson's sculptures.

42 salon

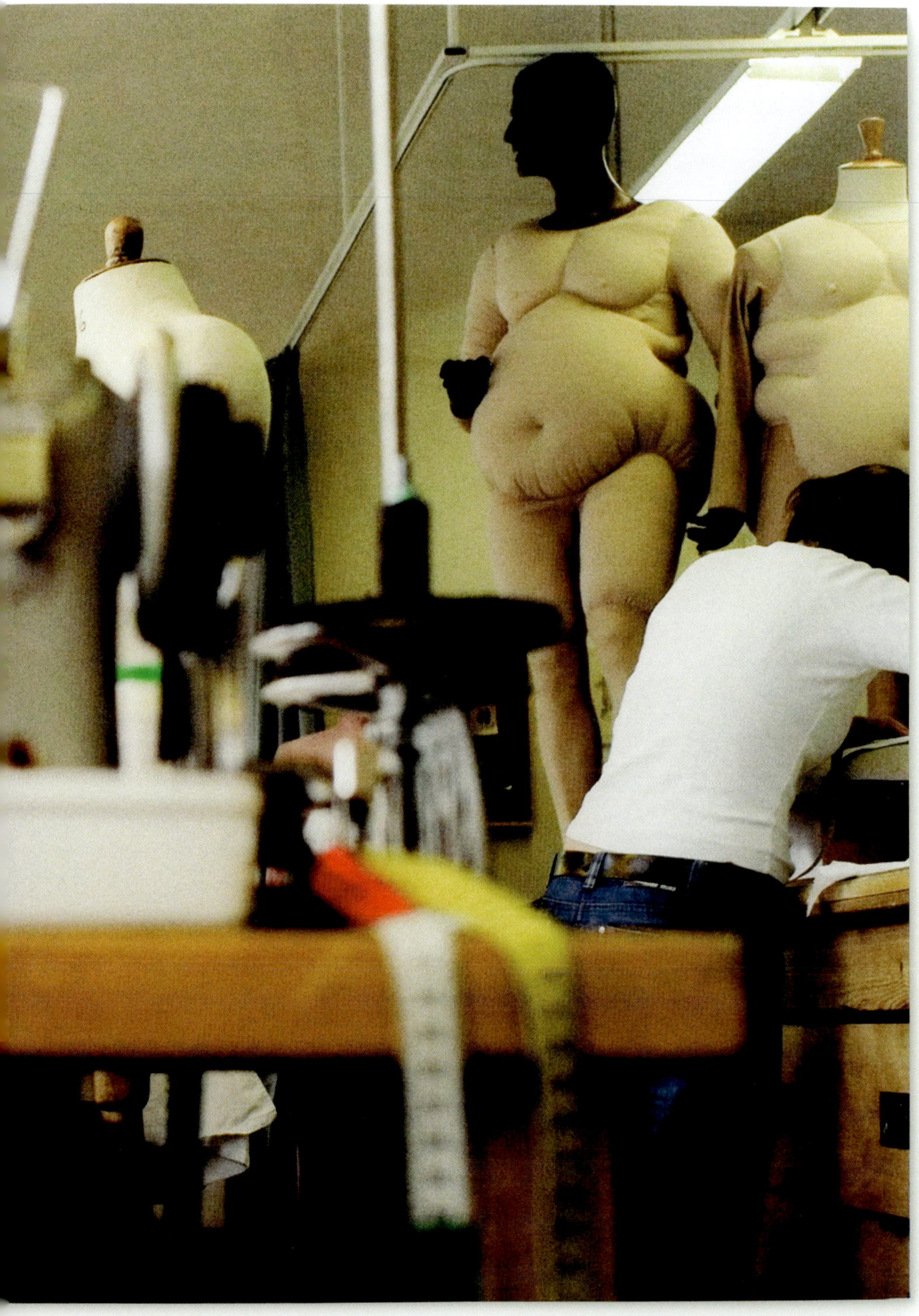

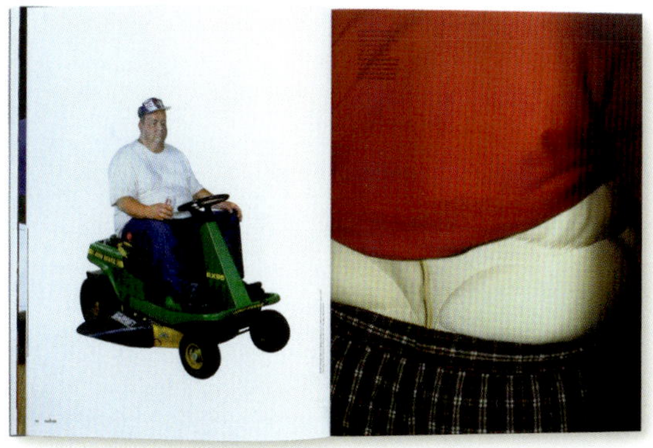

skyspace
SCHULE DES SEHENS
Photos: Torsten Braun
Als Oskar Kokoschka seine „Schule des Sehens" im Sommer 1953 auf der Festung Hohensalzburg einrichtete, hatte er die Vision, an einem außergewöhnlichen Ort einen Freiraum für Kreativität und Spontaneität zu schaffen. Diesen Anspruch hält die Internationale Sommerakademie für bildende Kunst bis zum heutigen Tag lebendig. Kokoschkas Leitgedanke, Wahrnehmung als spirituellen und emotionalen Akt zu begreifen, wird seit Dezember 2006 auch auf dem Mönchsberg zu einem synästhetischen Erlebnis: Sky Space, der begehbare Kunstraum des amerikanischen Lichtkünstlers James Turrell, ist ein buchstäblich himmlisches Sinneserlebnis.
SCHOOL OF SEEING
When Oskar Kokoschka opened his "School of Seeing" in the Hohensalzburg Fortress in the summer of 1953, he sought to create an open meeting area for creativity and spontaneity in an extraordinary place. Over the years, International Summer Academy of Fine Arts has maintained this standard of excellence. Since December 2006, visitors to the Mönchsberg have had the chance to experience Kokoschka's guiding principle, understanding perception as a spiritual and emotional act, as a synaesthetic phenomenon. Sky Space, the walk-in structure built by the American artist James Turrell, is literally a heavenly experience for the senses.

königin der kostüme
queen of costumes
DOROTHEA NICOLAI
LEITERIN KOSTÜMABTEILUNG
DIRECTOR OF COSTUME AND MAKE UP
Vita: Schneiderlehre in München, Praktikum am Bayrischen Staatsschauspiel, Fachhochschule für Gestaltung in Hamburg, Gewandmeisterin im Pariser Atelier für historische Kostüme & Haute Couture von Christian Lacroix, Kostüm-Assistentin in Brüssel. Rechte Hand von Herbert Wernicke. Ab 1995 Produktionsleitung an der Wiener Staatsoper, Lehrbeauftragte für Kostümgeschichte am Mozarteum und an der Münchner Akademie der Bildenden Künste. Mitglied im Kostüm-Komitee des International Council of Museums. Seit 1999 Kostümleitung bei den Salzburger Festspielen.

requiem

Jitzen oder nie.
SUPPORT LIVING ARTISTS
NO
TANGO ARGENTINO
salon

macher vor ort team on location
Zum Festspiel-Sommer 2007 treten Helga Rabl-Stadler und ihr kaufmännischer Direktor, Gerbert Schwaighofer, mit einer rundum neu besetzten künstlerischen Leitung an: Jürgen Flimm als Intendanten, Markus Hinterhäuser für Konzert und Thomas Oberender für Schauspiel. salon präsentiert das gesamte Team vor den neuen Spielstätten und blickt hinter die Kulissen der Festspielbühnen.
Text: Uta Gruenberger
Photos People: Gino Sprio
Photos Locations: Kris Scholz
salon
salon

Der Abschied fällt immer schwer
Saying goodbye is always hard
Klassische Musik erfüllte immer unser Haus". Reham Fayed lebt seit ihrem 13. Lebensjahr in Kairo, geboren wurde sie als Tochter eines ägyptischen Ingenieurs vor 27 Jahren in Saudi Arabien. „Ich liebte diese Musik. Besonders Mozart." Als ihr Vater sie zum Unterricht schickte, entschied sich Reham für die Querflöte. Seitdem übt sie jeden Tag mindestens fünf Stunden, auch wenn sie am selben Tag nochmals fünf Stunden Probe mit dem Cairo Opera Orchestra hat, für das sie schon mit 18 verpflichtet wurde. In Weimar, vor sieben Jahren, saß sie erstmals im West-Eastern Divan Orchestra. Unvergesslich. „Daniel Barenboim möchte perfekte Musik hören". Und sie liebt seine philosophischen Ausführungen, the way he speaks". Nie zuvor hatte sie mit einem Israeli Kontakt. Zuerst gab sie sich vorsichtig, „weil ich dachte, dass sie unsere Feinde sind. Doch dann habe ich gelernt, humanistisch zu denken". Vierzig Tage im Jahr sind Araber, Christen und Juden zusammen auf Tournee. „Der Abschied fällt immer schwer. Ich vermisse alle so sehr."
Mozart always comes along with Reham Fayed in her instrument case, when the Egyptian musician goes on an outing to the pyramids in the east of Cairo. Rehearsals, her daughter and husband fill her days. "I wish that classical music would become popular in our country."
salon

sommer nachts. traum mid summer nights' dream

simplicity

"Once upon a time there was a Dun Cow coming down along the maze and the Dun Cow that was down along the maze met a nicens little boy named baby Jim the bachelor."

Umberto Eco: *James Joyce – portrait of the young artist as a bachelor*, 1991

Umberto Eco's text on James Joyce – portrait of the young artist as a bachelor – was originally delivered as a contribution to a conference held at University College Dublin in 1991, to celebrate the centenary of Joyce's graduation as a Bachelor of Arts. The title is a pun on Joyce's own memoir, by someone who describes himself as a "playboy of the southern world!" After commenting on the different definitions of the word bachelor, all of which suggest an incomplete being (not yet married, not yet a doctor, not yet a knight, not yet a sexually successful adult dolphin), he asks the question of how immature Joyce was at that time, or whether the great ideas that were to produce *Ulysses* and *Finnegan's Wake* were already in motion.

He draws attention to Joyce's knowledge of and respect for Dante, who was to influence the construction of *Ulysses* as much as the Homeric model did. Dante wrote *The Divine Comedy* in Italian, but in an Italian for which he had himself created much of the vocabulary, borrowing from and regularising words from the many local dialects that Latin had diversified into after the Roman Empire, like the Tower of Babel, collapsed and fell. The principles for this development and the potential of a new, noble and poetic Italian were set out in his earlier doctoral thesis *De Vulgari Eloquentia*. Joyce too dissects, regrafts and rebuilds the English language in his search for expression, especially in *Finnegan's Wake*. Eco's comparison is thus an apt one. If his parallel attempt to connect Joyce with mediaeval attempts to rewrite the Irish language is more tenuous, he makes a strong case for Joyce's knowledge of the *Book of the Dun Cow*, one of the earliest works written in Irish, and of the magnificent imagery of the *Book of Kells*. His conclusion, that Joyce did indeed understand the task he had set himself, seems a fair one: at the end of the lecture Eco quotes Joyce's own words – those that appear at the head of this piece.

The language of a design, though largely written in visual signs, should perhaps also be such a hybrid. If the design is written in the designer's language only, it is unlikely to achieve what the client wants, and if it is written only in the client's language, then the client has perhaps learned nothing from the process, has not moved forward. A successful design should create a new language from the existing languages around it. Not in an attempt to be obscure, or difficult to read, indeed the contrary. The design and its language should share a complete clarity, appearing to have come into existence effortlessly. These are qualities Thomas Manss' designs possess amply. Looking at a range of them, one realises his keen sense of typography and his interest in the subtle use of colour, but individually they show a coherence and simplicity which belies the complexity of their making. But then isn't all good design complex analysis leading to simple execution?

Internationales Design Zentrum The Internationales Design Zentrum in Berlin sits at the hub of design life in the city, putting on exhibitions and hosting conferences, and acting as an information and advocacy centre for the design community and their clients. The invitation to design their annual report is an obvious accolade for a designer, the chance also to create a publication which will be seen – and looked at closely – by the chosen designer's peers in Berlin and beyond. The result often is that, as Thomas Manss puts it, "they design their little hearts out." When he was invited to design the 2003 Annual Report he decided a different, less indulgent approach was needed. He wrote to a large number of designers in Berlin, asking them to name something in the city that they particularly liked – a garden, or a bar or a shop or a restaurant or a building: whatever in the city touched their fancy. Having gathered a list together, he then asked the photographer Johanna Rübel to photograph each of them. The final design used the photographs as full page images facing the text of the report (which was in German and English, and so required some typographic verve to resolve successfully). A vertical note alongside each image identified the location and also who had selected it. So if one wanted to know which was Justus Oehler's favourite al fresco restaurant, Sebastian Turner's much-loved football ground or whatever, or who else liked the artist materials shop in Fuggerstraße, then one could find out. "I found places in Berlin I didn't know about myself," Thomas remarks, "and now if someone going to Berlin asks me to recommend places to see, I just give them a copy of the report."

What Thomas Manss achieved, which seems so simple in retrospect, was to substitute the individualistic language that others had applied to the report in the past, with a collaborative language, in which all could share (or all those who had contributed suggestions). This also was a subtle reminder of the relationship between the IDZ and the design community in Berlin at large, a shared publication for a shared facility.

Haben wir Angst vor Farbe?

Frostmale – Fotoarbeiten

IDZ Projekt 2003
Designzone Berlin. Kreativität als Standort- und Wirtschaftsfaktor

SIEMENS
mobile phone

IDZ Ausstellungen

Babetto ging von der kleinen in die große Dimension, vom Schmuck zur Inneneinrichtung. Sein Interesse für Möbeldesign ist stark mit seinem Interesse für Architektur verknüpft – eine Inspirationsquelle, die auch in seinen flächigen und oft geometrischen Schmuckstücken sichtbar wird. Konstruktivistisch puristisches Formenvokabular ist dabei charakteristisch für die beiden Gestaltungsbereiche Schmuck und Möbel. Doch unterliegen beide Werkbereiche jeweils anderen Gesetzmäßigkeiten: „Nur vergrößern geht nicht, dann verliert das Objekt seine Intensität." [...] „Was mich fasziniert ist die Designidee." (Giampaolo Babetto in „Gioielli di Cultura" herausgeben von Gli Ori Prato 2002).

Der Künstler versteht sich selbst als Klassiker. Sein Griff in die Kunstgeschichte – die Auseinandersetzung mit dem italienischen Renaissancemaler Jacobo Pontormo, Einflüsse der russischen Avantgarde oder des De Stijl, seine Affinität zu Mies van der Rohe, Norman Foster und den Künstlern der Minimal Art – sind für ihn visuelle Erlebnisse, Exkursionen, aus denen er seine eigene Formensprache entwickelt und autonome innovative Werke schafft.

Mit seinen Arbeiten ist Babetto nicht nur in Privatsammlungen vertreten, sondern auch in bedeutenden Museen wie zum Beispiel: Victoria and Albert Museum, London; Kunstgewerbemuseum, Berlin; Musée des Arts Decoratives, Paris; Museum für Kunst und Gewerbe, Hamburg; Museum of Art; Rhode Island School of Design, Providence, Rhode Island USA.

Mit Giampaolo Babettos Ausstellung „Jeweldesignartlink, Schmuck und Möbel 1968 bis 2003", setzt das IDZ Berlin die Tradition fort, Design aus anderen Ländern zu präsentieren. Speziell für die Ausstellung hat der Künstler eine „Edition Berlin" entworfen, eine Brosche in Gold mit blauem oder rotem Pigment.

In Babetto progression from the small to the large, from jewellery to furniture, the artist's interest for interior design reflects his love for architecture – a source of inspiration that in turn influences his flat and geometrical jewellery. Pure constructivist forms characterise all Babetto's creations, although jewellery and furniture follow entirely different rules: "You can't just enlarge, because it would reduce the object's intensity. [...] The idea of design fascinates me." (Giampaolo Babetto, Gioielli di Cultura, published by Gli Ori Prato, 2002)

Babetto considers himself a classical artist with historic roots. Jacobo da Pontormo and Italian Renaissance painting, influences from De Stijl and the Russian avant-garde of the 1920s, affinities with Mies van der Rohe, Norman Foster and minimalist artists: all these references are visual experiences for Babetto, like excursions from which his own style, his own innovative works arise.

Babetto's works are found in private collections and exhibited in major museums, including London's Victoria and Albert Museum, Berlin's Kunstgewerbemuseum, the Musée des Arts Décoratifs in Paris, Hamburg's Museum für Kunst und Gewerbe, and the Rhode Island School of Design.

The Giampaolo Babetto exhibition "Babetto Jeweldesignartlink – Jewellery and Furniture, 1968 – 2003" is a continuation of IDZ's tradition of presenting international design in Berlin. Especially for this exhibition the artist created a gold brooch with either blue or red pigments, titled Edition Berlin.

Neue Nationalgalerie, Potsdamer Straße 50, 10785 Berlin. Ein Lieblingsort von Uli Mayer-Johanssen. **A favourite place of Uli Mayer-Johanssen.**

vocabulary

Dans l'S, à une heure d'affluence. Un type dans les vingt-six ans, chapeau mou avec cordon remplaçant le ruban, cou trop long comme si on lui avait tiré dessus. Les gens descendent. Le type en question s'irrite contre un voisin. Il lui reproche de le bousculer chaque fois qu'il passe quelqu'un. Ton pleurnichard qui se veut méchant. Comme il voit une place libre, se précipite dessus.

Raymond Queneau: *Exercices de Style*, 1947

This banal text about chance encounters in a Paris bus and outside a station becomes in Raymond Queneau's hands the vehicle for a series of stylistic exercises, as he retells the same story in ninety-nine different ways. The retellings are in different tenses, and different styles, ranging from Alexandrine verse to an official letter to a press release to a series of questions or a series of exclamations. This kind of exercise is familiar from the tradition of the *ars rhetorica* of the early universities, though Queneau adds quite a number of newer varieties. And, as Umberto Eco discovered when he came to translate the book into Italian, quite a number of the traditional motifs are used in an allegorical way, and beyond that, despite the titles indicating the form or presentation, a large number of the examples in fact use several different methods superimposed into the final piece. In short, what sets itself out as a series of simple exercises is in fact a collection of subtle complexities that can be read in a variety of different ways.

The contemporary designer has a similar toolkit to give a stylistic gloss to a text, in the enormous range of different typefaces available. (A French publisher issued a special edition of *Exercices de Style* in which each entry was in a different and appropriate typeface.) Some typefaces were designed for supposedly purely technical purposes, such a Bell Centennial, created to make telephone directories less large and expensive to print, without losing legibility. Others arise from purely formal considerations, such as the Bauhaus faces created exclusively from straight lines and perfect circles. But whatever their origins, all typefaces carry an emotional and stylistic charge, which derives both from their visual form and how this is interpreted by the user or reader, and from how they have been used in history (for example, traditional German 'black-letter' typefaces disappeared from use in the late 1940s and 1950s, not because they were old-fashioned – though they are sometimes difficult to read for unaccustomed eyes – but because their use had, for a time, been enthusiastically promoted by the Nazis).

The subtleties of typography are part of the wider range of variations a designer can bring to publications: size and proportion of the page, the relation of text to illustration, the use of colour both as a background and in type, the weight and surface finish of the paper are among the others. The designer's task is to produce a design that is both apt and different: appropriate for the subject and also different from other publications in the same field. This is an approach that does not lend itself to a house style: it requires individual attention and precise decisions, so that all shades of meaning can be evaluated. Thomas Manss' experience and range in working for publication covers both books and magazines across a variety of fields, each one an individual solution.

ZAHA HADID
PROJECTS DOCUMENTATION
Thames & Hudson
ZAHA HADID
MAJOR AND RECENT WORKS
Thames & Hudson
ZAHA
HADID
Thames & Hudson
ZAHA HADID
PROCESS: SKETCHES AND DRAWINGS

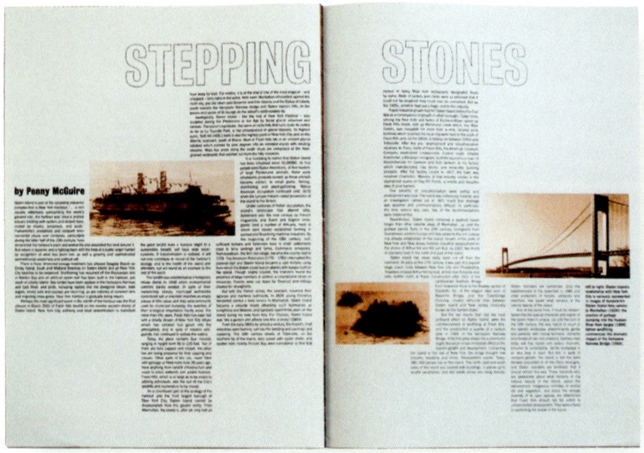

LETTER FROM STATEN

below: Extracts from video-taped interviews with Staten Island residents by Mark van S., November 2001.

'The farms are gone. The farmers couldn't pay for the taxes on the land because it was so valuable as real estate land. In fact it was probably the bridge coming that ended the last farm.'

'The people of Staten Island have had an overdose of ugly things and to make a great beautiful sweep of a landscape would be a healing tonic to this kind of harsh, urban environment.'

'I heard that they were going to turn it into a park, but I don't know if that's happening. There's been a lot of developing recently, like townhouses are popping up all over and a lot of historic houses have been torn down. We've had a lot of developing and I think we need the trees back.'

ISLAND

by Tamara Coombs

When it was open country it was nothing but farms and farm land around. I am a cowboy, everything about me is cowboy. sorta stand out like a sore thumb in Manhattan, Brooklyn and Queens.'

When Fresh Kills Sanitary Landfill opened in 1948, Robert Moses assured a concerned letter writer that the landfill would close within three years. Moses missed the mark by half a century. To those who had never been closer to the borough than the decks of the Staten Island Ferry, the landfill became the Island's most famous feature. To those who lived on Staten Island, the landfill was an unpleasant reality and a daily reminder of the borough's place in the City's pecking order. The stench was always perceptible to those nearby, generally worse in the summer and often carried inland on the prevailing westerly winds. On top of the engineered hills, flocks of raucous seagulls fought over exposed garbage. Two rows of tall wire fences caught much of the windborne trash, but flimsy plastic bags escaped and festooned the branches of trees and shrubs.

All of this was there to be seen and smelled as Staten Islanders drove the West Shore Expressway (windows up) or went shopping at the Staten Island Mall or to the movies at the Dump. For some history-minded Staten Islanders, the Dump fitted into a pattern of using the Island with its low population and undeveloped land as the site for facilities others didn't want. This pattern began with the Quarantine Hospital (1799 – 1858) and continued with the Tuberculosis Sanatorium at Sea View (1905 – 1961) and the infamous Willowbrook State School (1952 – 1988).

The lack of vision and long term planning for the borough was demonstrated with the completion of the Verrazano-Narrows Bridge in 1964. Although some may be nostalgic for a perfect past that never existed, there is no denying that much of Staten Island's peaceful South Shore was transformed into tacky subdivisions and strip malls. Today, traffic jams and inadequate public transportation have given Staten Islanders the dubious distinction of having the longest median commutes in the United States. Surprisingly, thanks to dedicated citizens and true public servants, much of value remains including communities with tree-lined streets and architectural character, a Greenbelt with miles of trails winding through pine oak woodlands and a waterfront of enormous potential. Still, despite its assets, Staten Island has often been viewed disdainfully by the rest of New York City. It has been seen as a boring blue-collar suburb, full of cops and firemen, notable only for its landfill. As one T-shirt put it, Staten Island: World's Largest Dump.

The Fresh Kills Landfill closed in March of 2001. It reopened September 13, 2001 to accept tons of World Trade Center rubble mixed with the remains of thousands of victims, including hundreds of policemen and firemen. Over 20% of the uniformed personnel lost at the WTC lived on Staten Island. Retired firemen and policemen volunteered for the terrible duty of sorting through debris at Fresh Kills in search of something identifiable to give to grieving families. The unsung men and women of the FDNY and NYPD have metamorphosed into heroes. A despised landfill has been transformed into hallowed ground.

It would be hard to overemphasise the opportunity Fresh Kills presents. From a borough that has been treated as a working class joke could come a world-class example of how to reclaim the planet's largest landfill and turn it into a self-sustaining landscape environment of great beauty and usefulness. For Staten Islanders, a source of anger and sorrow could become a source of pride and pleasure. It would be a welcome assurance that we can begin to repair what we have despoiled; a note of optimism in a world in need of it.

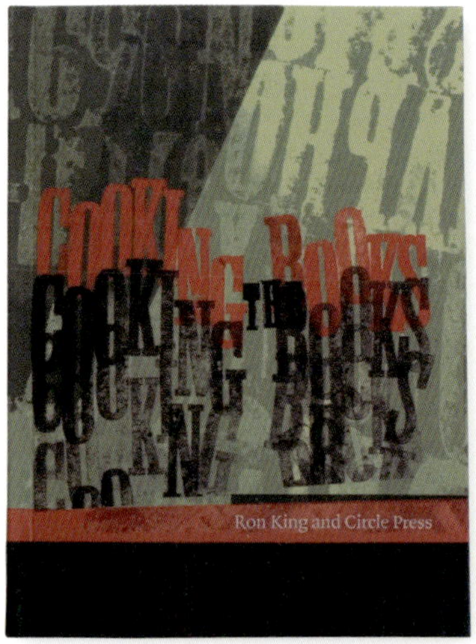
Ron King and Circle Press

RIVERLINES

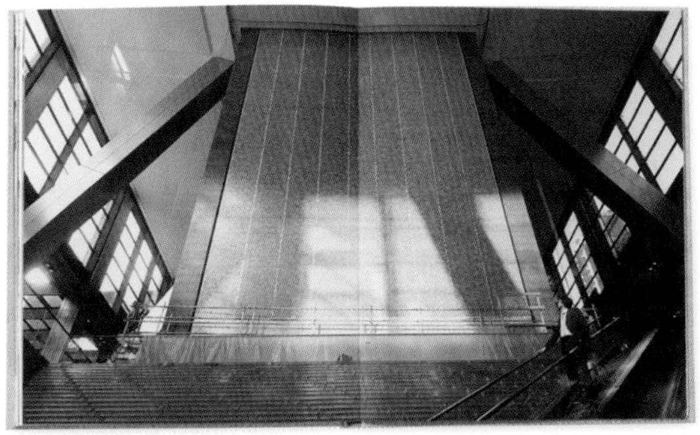
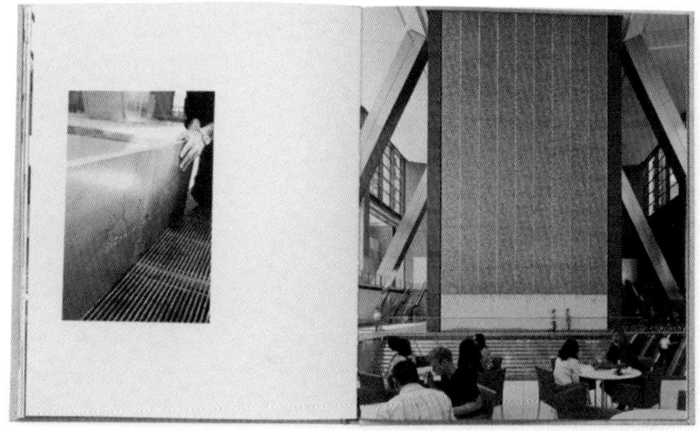

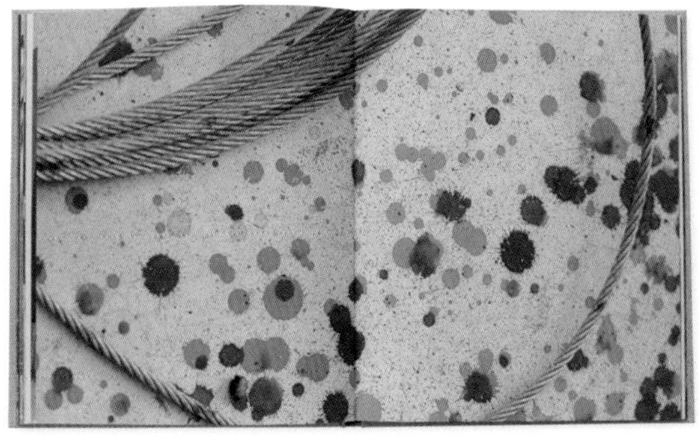

wit

***Aga Saga*: term applied, usually perjoratively, to a sub-genre of popular literature (GENRE) set in a semi-rural area and centred on the domestic and emotional entanglements of affluent middle-class characters. It derives from the proprietary name of a type of stove which has come to symbolise a wealthy cosiness.**

David Macey: *Dictionary of Critical Theory*, 2000

Putting the *Aga Saga* into a *Dictionary of Critical Theory* suggests a somewhat wayward spirit on the part of the author, David Macey, and indeed the book is filled with entertaining asides and comments on all sorts of subjects. The extensive cross-referencing allows the reader to follow threads of ideas, for example creating a history of narrative (although narrative is not a topic as such in the dictionary).

Narrative is a concept that dominated (even if from the background) artistic and literary discourse from the world of the ancient Greeks until the rise of Modernism in the early twentieth century. The Modernists' concern with function and their hostility to decoration closed off some of the usual routes into narrative, in design terms, but the waning of Modernist influence has allowed narrative to re-emerge in new forms. (The work of such very different artists as Joan Fontcuberta and Tracey Emin can be seen as each having a narrative core, for example, as does the new genre of time-based art, whether through video or performance.) So I would suggest that there is enough substance in an idea of narrative (or, rather, different ideas of narrative) to sustain a discussion about design that sees narrative or storytelling as one of its main agents, that can be cross-referenced to the literary experience of recent years or before. This is what, on perhaps a ludic level, this book seeks to do, by seeking to build parallels between different kinds of literary discourse and the practice of design.

Narrative can be seen, in a sense, as the victor in the postModern revolution. Not the closed narratives of the pre-Modern period, but an active, open and engaged narrative that connects consumers to design and through that individuals to society. This book uses narrative because, I believe, the work of Thomas Manss & Company engages with the postModern narrative in interesting, valid and successful ways, as well as bringing to that process the traditional and modern skills of graphic design.

My recollections of meetings with Thomas Manss over the last ten years or more are of much laughter and no less delight. But to suggest that a designer is light-hearted does not imply that he is also light-headed. Walk into Thomas Manss' present offices, just off the City Road in London, and the immediate impression is of an ordered, organised space. The front door seems to stand at the apex of the office, with its flight of steps to the street. More than once his colleagues have answered my ring at the door with an out-thrust package, assuming a pedestrian author must be a fleet motorcycle messenger. Once admitted the visitor encounters an initial welcoming space with seating and a set of shelving (yes, Vitsœ 606) displaying their recent literature work. For a moment the trick of triangularity lingers as beyond that the space widens out for a dozen or so workstations, with a library and file stack on the right hand wall, and at the end a circular space which functions as a conference room. Thomas Manss' own office is simply a desk against the end wall of what is in fact a rectolinear space. It is an uncluttered space: what adds touches of colour are the fragments of designs seen on desks and screens, as well as Thomas Manss' Ozwald Boateng suits, of course. (He wore one for a presentation to the UK Carnival Arts Trust and was asked if he had bought it for the occasion: "no," he replied, "I always wear them." He got the commission to design their identity and website.)

Colour and typography are the staples of the graphic designer's trade: to say that Thomas Manss and his team are masters of these is to state the obvious. There are, however, three other qualities which are more specific to the agency: its international outlook, its use of wit, and, most importantly, the concept of narrative.

Quite apart from Thomas Manss' German background and practice and British experience, the agency has staff from England, Germany, France, Italy, America and Russia. This has practical advantages: the Russian designer helped bring in one of their first projects in Moscow with the ProfMedia group: their logo is being rolled out later this year. More importantly, this means that a whole range of different cultural perspectives and contexts can be brought into use on any project, whether for a Spanish hotel, an American business or a British creative arts consultancy. This facility to look at a brief in different ways from the start is a valuable aspect of the design process.

Wit is a quality that should be used sparingly: it is nonetheless a valuable tool. The idea is not to have the end-user break into fits of giggles when seeing a logo, which would be complete overkill, but rather of raising the sort of quiet smile one has when one sees someone one knows, or shares a private memory in a wider conversation. The logo for copywriter Val Taylor looks like a pen-nib: closer inspection shows it is a V and an B superimposed. Once one has worked out the trick, it stays in the mind. The monogram for IT consultants Transformal combines the Initials T and F to illustrate the firm's offer: the key to IT. Again, the idea sticks in the head. It clicks!

Narrative is the real key: "if you want to create a memorable design, you have to start with a thought worth remembering," Thomas Manss has observed. And often what is most memorable is a story. This book has tried to use a number of different forms of narratives or of narrations to illustrate how such forms integrate with different aspects of design and how that integration works in the case of projects created by Thomas Manss & Company. These are specifics: there are other instances, as the use of narrative in defining, developing and creating a brief from a design is a central part of their work: it's part of their magic.

Prof-Media Management – Media Holding

UK Centre for Carnival Arts

The Henry Lydiate Partnership – Business Consultants
Brainshell – Patent Agency
Stern- und Kreisschiffahrt Berlin – Cruises

Ability International – Charity
LogistikNetz Berlin-Brandenburg – Logistics Network
Lienke – Recycling

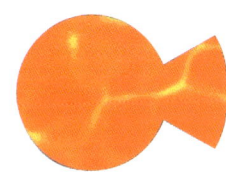

HighTechCenter Babelsberg – Post Production
Netzwerk Weiße Biotechnologie Berlin-Brandenburg
Preussische Schlösser & Gärten – Museum Shops

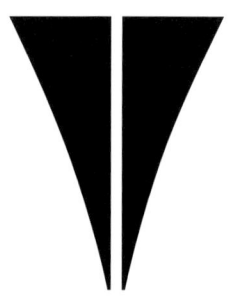

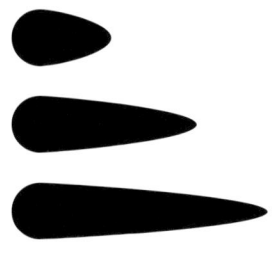

Unicef – Children's Charity
192 – Bar and Restaurant
Nautilus – Loudspeaker Technology

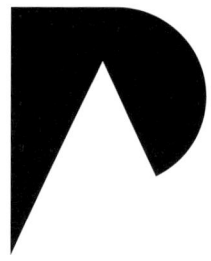

Prism – Loudspeaker Technology
Val Taylor – Copywriter
Nautilus – Loudspeaker Technology

LA21 – Landscape Architects
Booktailor – Valentine Promotion
Flowport – Loudspeaker Technology

VCC Perfect Pictures – Post Production

Depken & Partners – Management Consultants
Transformal – IT Consultants
Horne – Civil Engineering

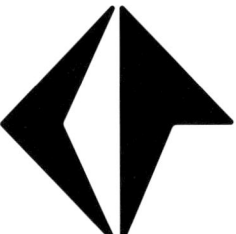

Axentum – Asset Management
Metamorphosis – Image Consultants
Lean Alliance – Management Consultants

Land Brandenburg – Technology Marketing
Atlantic Energy – Energy Supplier
First Source – Purchasing Club

Technologie Stiftung Brandenburg – Technology Marketing
Tim Wood – Furniture Designer
Datatrain – IT Consultants

ERP – Enterprise Resource Planning
Schober Eiche – Management Consultants
eComm – Competence Centre

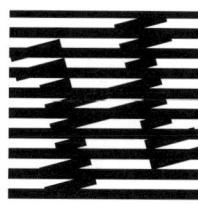

Hotel Arts Barcelona – Luxury Hotel
Lill – Property Lawyers
HAL– Computer Merchants

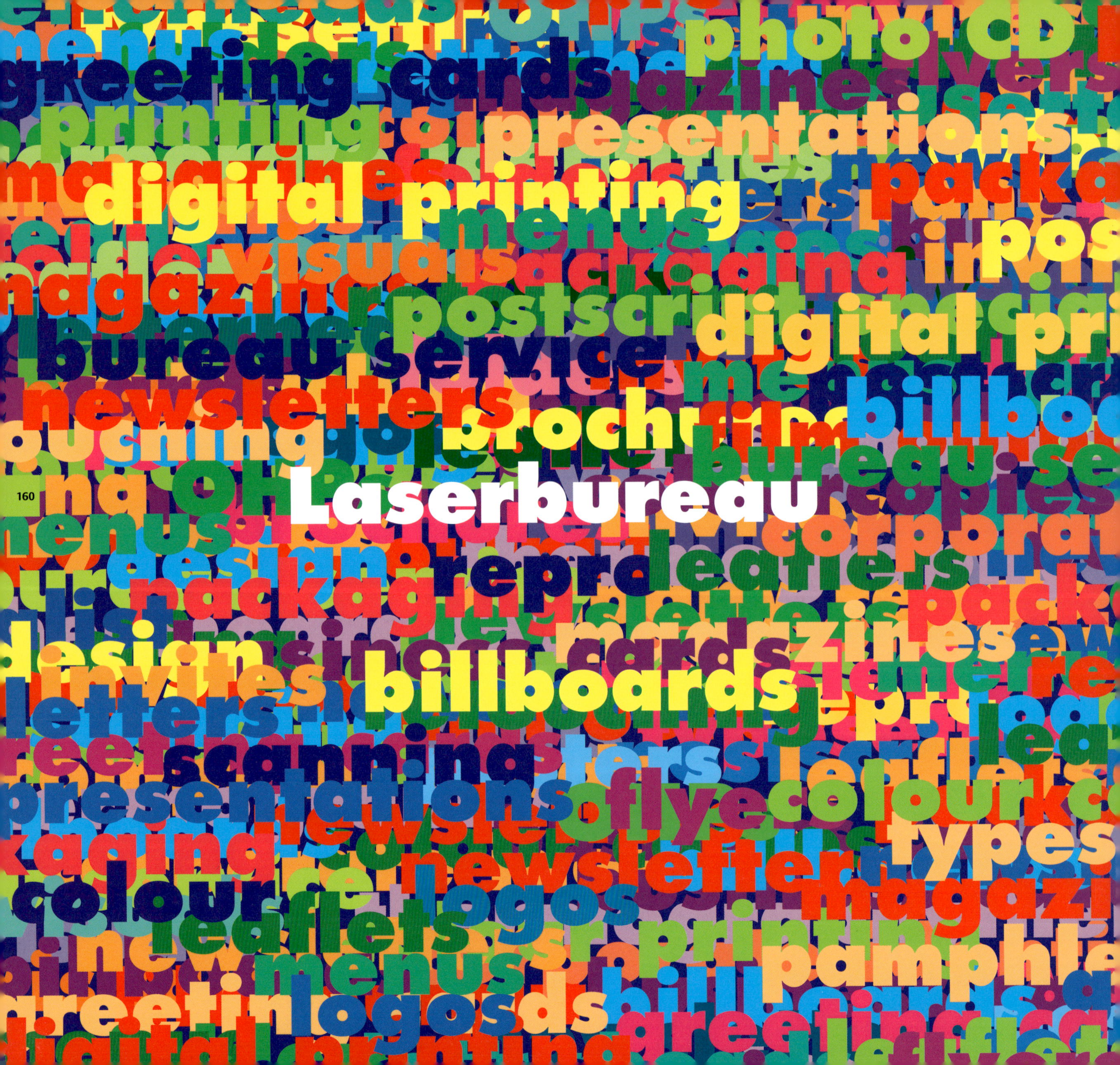
photo CD
greeting cards
presentations
digital printing
menus
postscript
digital
bureau service
newsletters
brochu
billbo
Laserbureau
repro
leaflets
billboards
scanning
presentations
colour
types
newsletter
logos
leaflets
pamphl
menus
logos

micr°pelt

K.R.P.R

CLASSE

mindseye

B&W

ROTEL®

VITSŒ

breeze

QUARTERLY

cutcost.com

vertixx.

openstudio

Micropelt – Thermo Electrics
Mindseye – Architectural Lighting
Vitsœ – Shelving
Cutcost – Office Supplies

Katie Rosenberg Public Relations – PR
Bowers & Wilkins – Loudspeakers
Breeze – Energy Consultants
Vertixx – Internet Distributor

Classé – Audio Electronics
Rotel – Audio Electronics
Art Quarterly – Art Fund Magazine
Openstudio – Architects

Hopkins Architects

ivorypress

SQUIRE AND PARTNERS

MATILDI & PARTNERS

GRIMSHAW

Ludvigsemøbler

Knippers Helbig

bachchor

markbraun architekten

Contemporary | Art | Society

Seidel & Friends

interActiva

John McAslan + Partners

PRINGLE | RICHARDS | SHARRATT | ARCHITECTS

CANNONBRIDGEHOUSE

Hopkins Architects – Architects
Matildi & Partners – Engineers
Knippers Helbig – Engineers
Contemporary Art Society – Art Consultants
John McAslan + Partners – Architects

Ivory Press – Publishers
Grimshaw – Architects
Bachchor – Choir
Seidel & Friends – IT Journalists
Pringle Richards Sharratt – Architects

Squire and Partners – Architects
Ludvigsen Møbler – Furniture Manufacturers
Mark Braun – Architects
InterActiva – Event
Cannon Bridge House – Property Development

Knaben chor gütersloh

Knabenchor Gütersloh – Boys Choir

FHP:->

PATRICK HEIDE CONTEMPORARY ART

Selected clients

Abakus Wertkontor AG
Ability International
Alain de Botton
Alan Baxter & Associates
Arjo Wiggins Fine Papers
Artemedia
Artnet
Arup
Atene
Atlantic Electric & Gas
Aura
Axentum Asset Managment

Bachchor Gütersloh
BBX
Benjamin Franklin Center
Berliner Verkehrsbetriebe
Berlin Transport
BKS Consult
Black Dog Publishing
BMS Dr. Stenner
Booktailor
Bowers & Wilkins
Brainshell
Breeze
British Museum
Business Link

Calmann & King
Cannon Bridge House
Chapmann & Hall
Chelsfield
Circle Press
Classé
Contemporary Art Society
Cost Reduction Management
Coverdale
Cutcost.com

Danny Lane
Datatrain
Deborah Richardson
Deloitte
Depken & Partner
Deutsches Human Genom Projekt
Divers
Domus AG
Dr. Dirk Jung
Dr. Eckhard Löhde

East London Business Association
East London Partnership
Ecosse
ERA
Ettwein Bridges Architects

Fabriano
Fachhochschule Potsdam
First Source
Focus Business Center
Foster + Partners
Freie Universität Berlin

Gainwest
GenProfile AG
GeoForschungsZentrum Potsdam
Gesellschaft für Abfallbeseitigung Oberhavel
Gitanes
Grimshaw
Groucho Club
Growtth

HAL
HighTechCenter Babelsberg
Hopkins Architects
Horne
Hotel Arts Barcelona

IBIS
Ilka Martin
Immobilienart
IMP
Imperial College
Indigo
Industry Lead Body for Design
Innovationsmarkt Brandenburg
Institute of Directors
InterActiva
Internationales Design Zentrum Berlin
Ivory Press

James Stirling Foundation
Jennie Moncur
John McAslan + Partners

Katie Rosenberg Public Relations
Klett Verlag
Knabenchor Gütersloh
Knippers Helbig
Kulturverlag Polzer

LA21
Laserbureau
Laurence King Publishing
Le Walk
Lienke
Lill Rechtsanwälte
LogistikNetz Berlin-Brandenburg
London Records
L&R Productions

MacIntyre
Mark Braun Architekten
Mark Finley Consult
Mark Richards
Martin Markcrow
Matildi & Partners
McDonald Egan
Mediacs
Media.net Berlin-Brandenburg
Meoclinic
Messe Berlin Reed
Metamorphosis
Michael Wilford & Partners
MicroPelt
Mies van der Rohe Haus
Mindseye
Moveo
MTB
MuseumsShop Concept
Myryad

National Portrait Gallery
National Art Collections Fund
Netzwerk Weiße Biotechnologie

Oberhavel Holding
Oberhavel Verkehrsgesellschaft
Ogilvy Adams & Rinehart
Openstudio
Oyuna Cashmere

Parliamentary Works Directorate
Patrick Heide Contemporary Art
Pepperl + Fuchs
Phaidon Press
Phonetracker
Portsmouth Naval Dockyard
Preussische Schlösser und Gärten
Prestel
Pringle Richards Sharratt Architects
Prof. Boltz
Prof-Media Management

Rail Industry Training Council
Redstone Press
Ron King
Rotary Club
Rotel
Royal Institute of British Architects

Sarah Hutchins
Sarah Medway
Schober Eiche
Scholz & Friends
Schüring & Andreas
Seidel & Friends
Sennheiser
Shilla Hotels
Smart Process
Spanish Heritage
Spirit of Creation
Squire and Partners
Stadt Gütersloh
Stadtwerke Bielefeld
Stemme
Stern und Kreisschiffahrt Berlin
Symtec

T+I Consult
Taschen Verlag
Tate Britain
Tate Modern
Technische Fachhochschule Wildau
Technologie Stiftung Brandenburg
TechnologieAllianz
The British Foreign & Commonwealth Office
The Islamic Art Society
Thomson ELT
T.IN.A. Brandenburg
T.I.NET. Brandenburg
Tim Wood Furniture
Trias Consult

UK Centre for Carnival Arts

Val Taylor
VBB Verkehrsverbund Berlin Brandenburg
VCC Perfect Pictures
Vertixx
Vitsœ

Wallraf-Richartz Museum
Wassily
Wirtschaftsförderungsgesellschaft Oberhavel
Whitegoods
Wigmore Hall
Wilkinson Eyre Architects
Wirtschaftsministerium Land Brandenburg
Wirtschaftsförderung Brandenburg
Witte, Weller & Partner
Worshipful Company of
Information Technologists

Zaha Hadid Architects
ZukunftsAgentur Brandenburg

192 Bar and Restaurant

Bibliographic information published by Die Deutsche Nationalbibliothek
Die Deutsche Nationalbibliothek lists this publication in the Deutsche Nationalbibliografie; detailed bibliographic data are available in the Internet at http://dnb.d-nb.de.

Texts: Conway Lloyd Morgan
Editing: Petra Kiedaisch, Jo Stead
Design: Thomas Manss & Company
Printing: Leibfarth & Schwarz, Dettingen/Erms

ISBN: 978-3-89986-089-4

Printed in Germany

www.avedition.com

Photos and illustrations

Frank Cremers 20-21
Steve Double 22
Große 8, 15
Devin de Haven 22
Ken Kirkwood 90-91
Wolfgang Korall 35
Gregory Krum 100
Live from Abbey Road 25
Manfred Michel 28-29, 35-36
Manuela Montella 108
MIT Media Lab 24
Steve Rees 10-11, 48
Johanna Rübel 130-136
Hans Scherhaufer 47
Mark Sheldon 92-93, 101
Phil Sills 16-19
Gino Sprio 22
York Tillyer (Real World) 26
Vitsœ 101
Nigel Young 70-71, 74-75
Stills from *Bowers & Wilkins – A Sound Experience* 12-13
All other photographs: Thomas Manss & Company

Special thanks to

Nick Evans
Olga Gusarova
Andreas Lerchner
Joana Niemeyer
Vita Piccolomini
Domitille Pouy
Stefan Schroschk
Karl Shanahan
Sandra Zellmer

Thomas Manss & Company

3 Nile Street
London N1 7LX
T +44 20 7251 7777
F +44 20 7251 7778
E uk@manss.com

Sybelstraße 68
10629 Berlin
T +49 30 8867 7071
F +49 30 8867 7072
E de@manss.com

www.manss.com